EASY SCIENCE for EVERYONE

Science ideas help us understand our world!
Everyday, Science flows all around us.
Science is in the oomph that powers our
planet, our everyday objects and ourselves.

The Indē Ed Project Charitable
Organization's mission is to share
Easy science multimedia content.
To date, million of our ebooks and videos
are downloaded by people in 60 Countries.

Indē Ed Project
Charitable Org'n

Easy Ideas 1

Airplanes 2

Cars 3

Computers 4

Smartphones 5

Food 6

Nature 7

Space 8

Light 9

AI 10

Everyday Objects Bonus

Easy Ideas From Concepts to Critical Thinking 1	**Airplanes** From Four Forces to Flights 2	**Cars** From Actions to Autos 3
Computers From Digital to Data 4	**Smartphones** From Calls to Global Connects 5	**Food** From Eats to Energies 6
Nature From Atoms to All Life 7	**Space** From Elements to Us 8	**Light** From Suns to Sapiens 9
AI From Machine Muscles to Minds 10	**STEM-Zen Program** From Empty to Science EnLights	**Everyday Objects** From Ideas to Daily Items Bonus

Easy Science
— with Ideas, Flows & Powers

Table of Contents

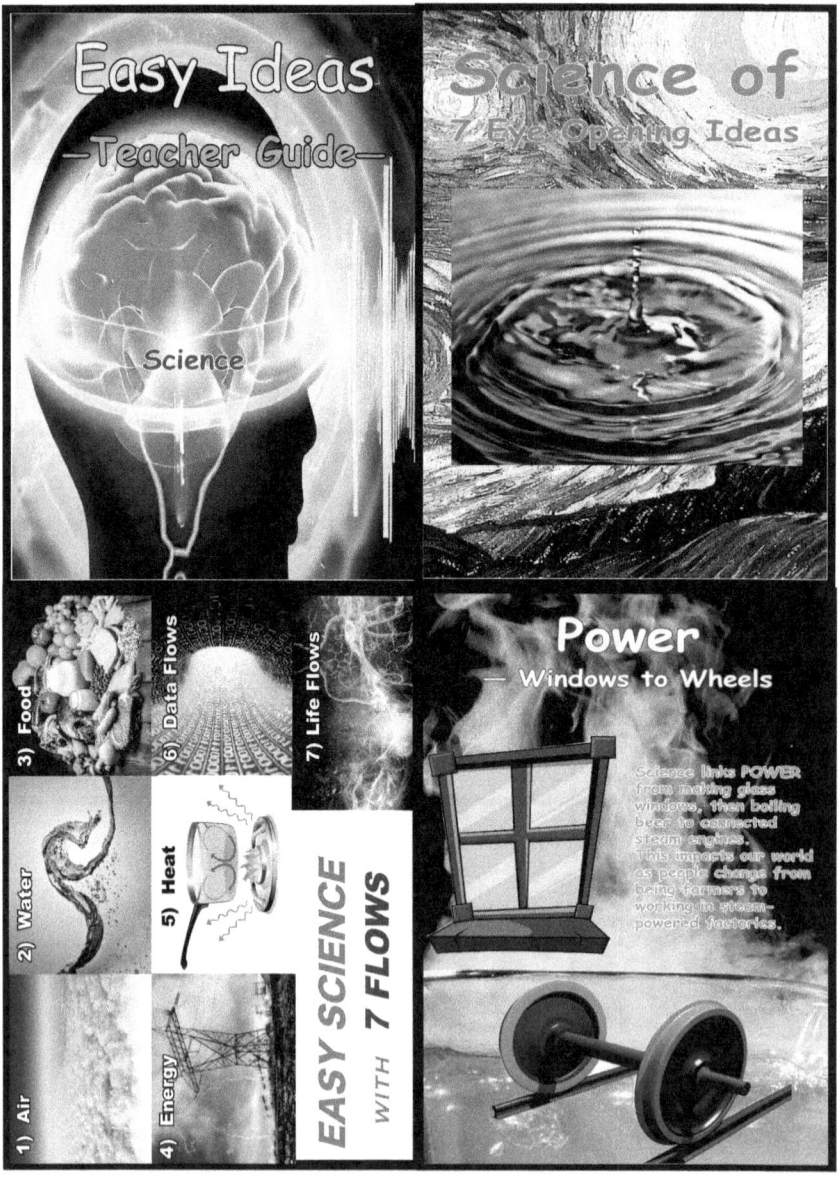

Easy Ideas
—Teacher Guide—

Science

Science of
7 Eye-Opening Ideas

3) Food

6) Data Flows

7) Life Flows

2) Water

5) Heat

1) Air

4) Energy

EASY SCIENCE WITH 7 FLOWS

Power
— Windows to Wheels

Science links POWER from making glass windows, then boiling beer to connected steam engines. This impacts our world as people change from being farmers to working in steam-powered factories.

1) Easy Ideas

—Teacher Guide—

Science

Douglas J. Alford
Sally Kimangu

STEM-Zen Program

1) Easy Ideas

We start our science quest with 7 ideas from atoms to apples that open our minds.

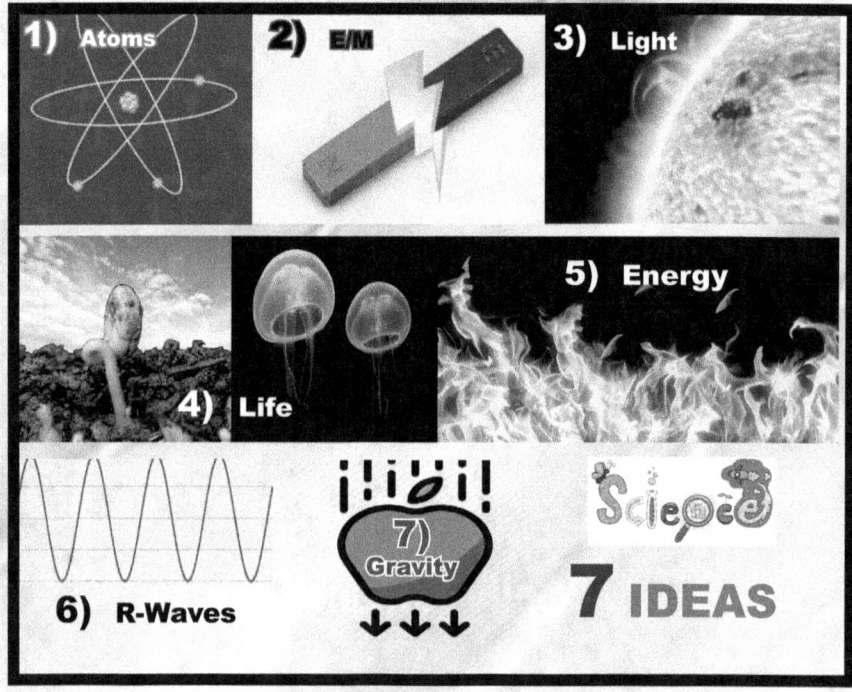

1) Atoms
2) E/M
3) Light
4) Life
5) Energy
6) R-Waves
7) Gravity

7 IDEAS

Easy Ideas
Table of Contents

1) ce of
Seven Opening Ideas
STEM-Zen Program

A) <u>Read</u> — ONE PAGER — Overview

Read

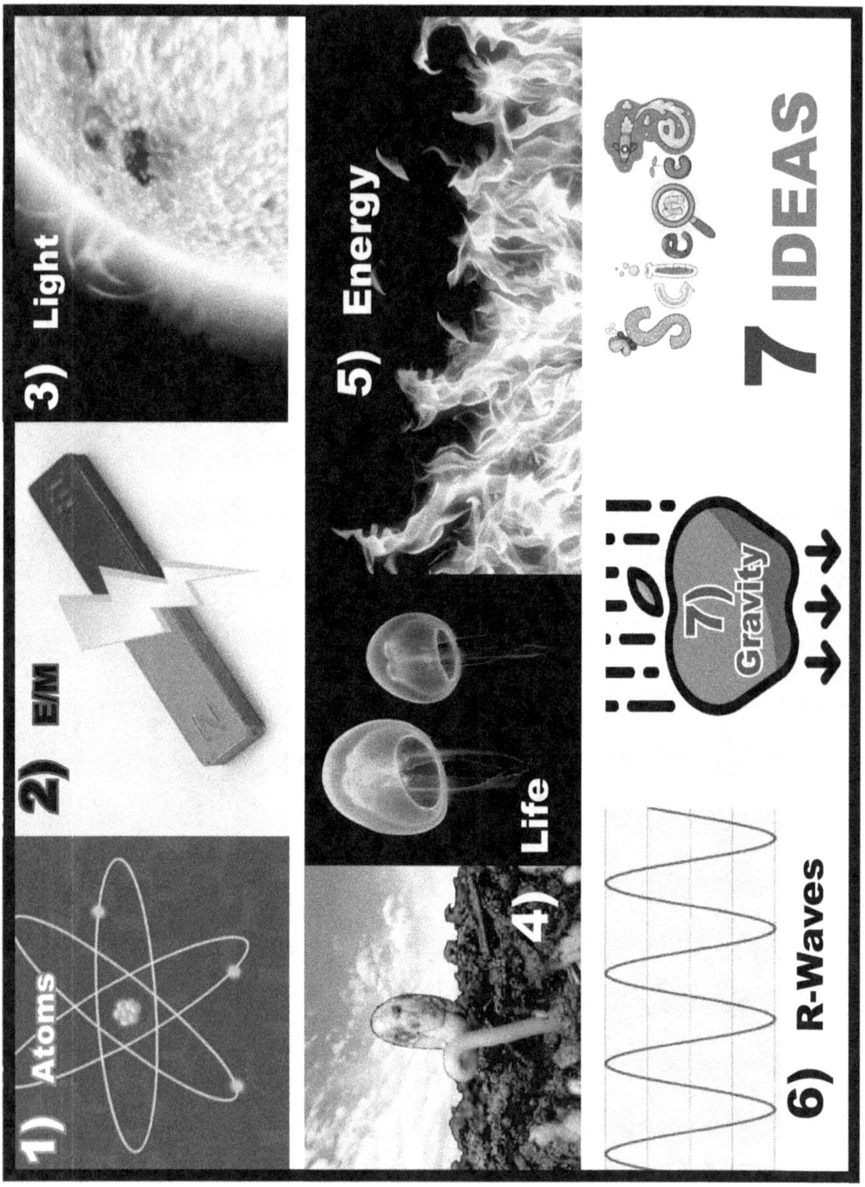
1) Atoms
2) E/M
3) Light
4) Life
5) Energy
6) R-Waves
7) Gravity
Science
7 IDEAS

B) Roadmap — Easy Ideas

Purpose: Show how science is important to our lives. Learn a little about wide subjects that are foundations for understanding science. We use this knowledge in future lessons too.

1) Everything on Earth is made of atoms.

2) Electricity and magnetism are related.
Batteries make electricity.
Outlet electricity powers our homes.
Electricity flows and makes magnets.
Magnets have two sides (poles).
Wires turn near magnets to make electricity.

3) Light can bend (refract), bounce (reflect) and beam (shine).

4) Life changes. Life is also the interactions of matter and energy.

5) Energy changes form. We use different types of energy to power electric motors, gas cars and jet-fueled airplanes.

6) Radio waves are useful. Radio waves connect people around the world with smartphones and the Internet. Smartphones use radio waves to make calls.

7) Gravity attracts. Gravity pulls apples, objects and us towards the Earth.

C) <u>Refresh</u> KEY CONCEPTS
— Energy Matters to Us

Energy matters. At the Big Bang, large amounts of energy become cosmic bits of matter that later become countless Suns and planets. Einstein's famous equation shows us that energy and matter are related. That is, energy becomes matter and also matter can change back into energy again. Welcome to the true story of science with energy and atoms on cosmic and quantum level adventures.

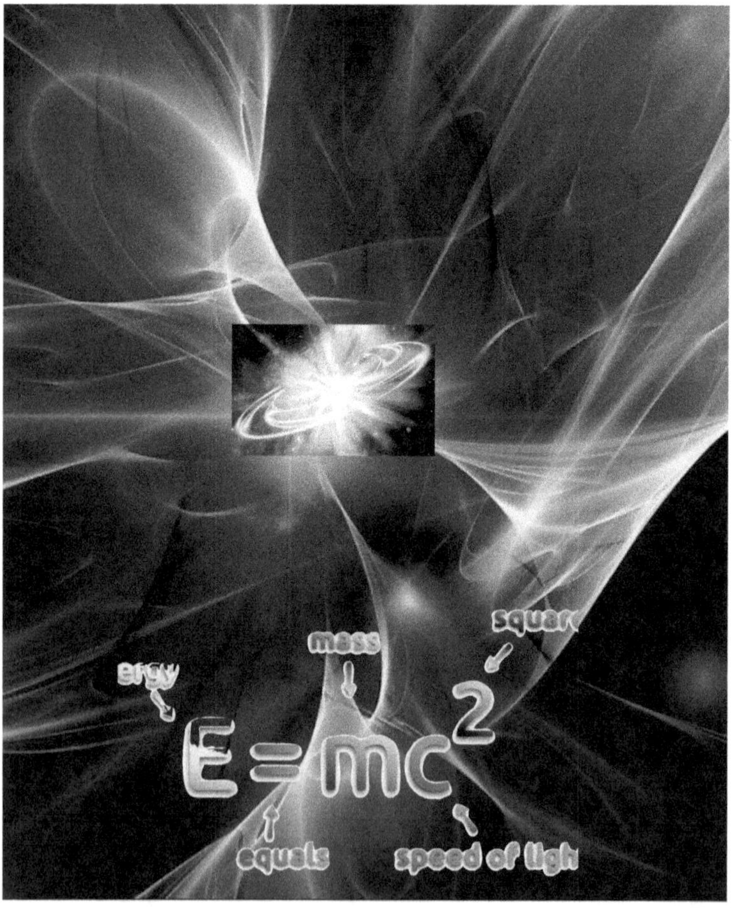

4

KEY CONCEPTS — 1) Atoms

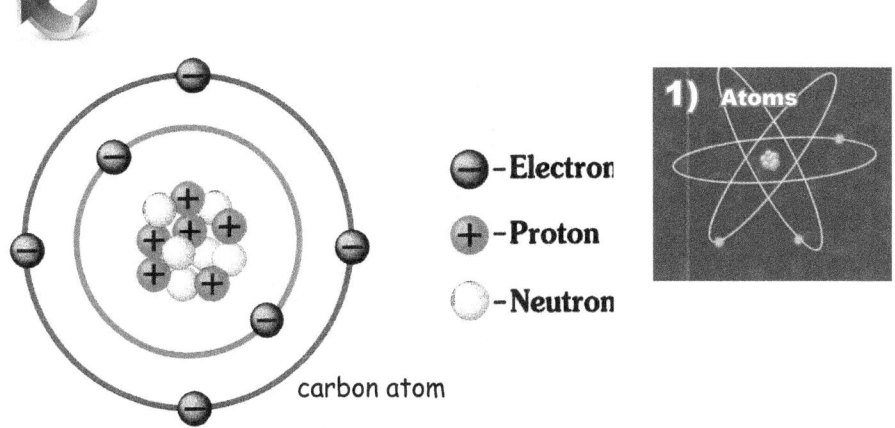

- **⊖** - Electron
- **⊕** - Proton
- **○** - Neutron

carbon atom

Our world is made of about 100 different types of atoms called elements. Atoms are made of positive protons in the center (nucleus) and negative electrons in orbit. Different numbers of protons and electrons make different atoms. Neutrons have no charge but may help keep the nucleus together.

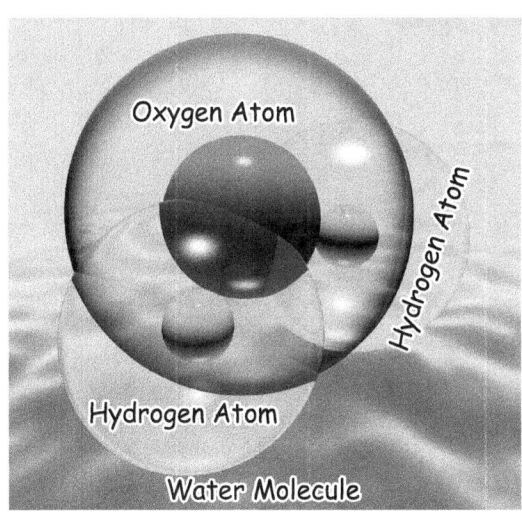

Oxygen Atom

Hydrogen Atom

Hydrogen Atom

Water Molecule

Atoms join together to make molecules. One oxygen atom and two hydrogen atoms make one water molecule ($H2O$).

2) Electricity

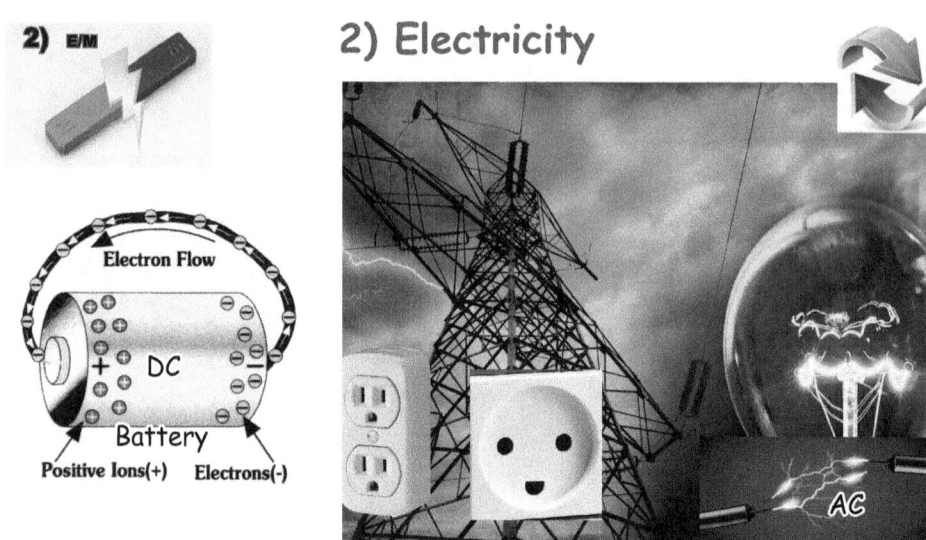

Electricity is energy made by flowing electrons. Direct current (DC) flows in batteries. Alternating current (AC) flows out of outlets.

Magnetism

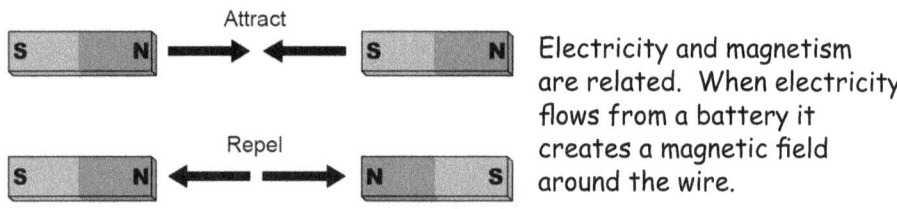

Electricity and magnetism are related. When electricity flows from a battery it creates a magnetic field around the wire.

Electromagnetism

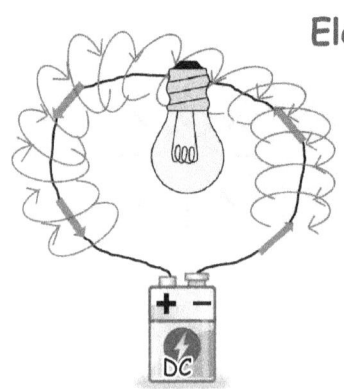

When electricity flows from a battery it creates a magnetic field around the wire.

Teacher - Seven Ideas

3) Light

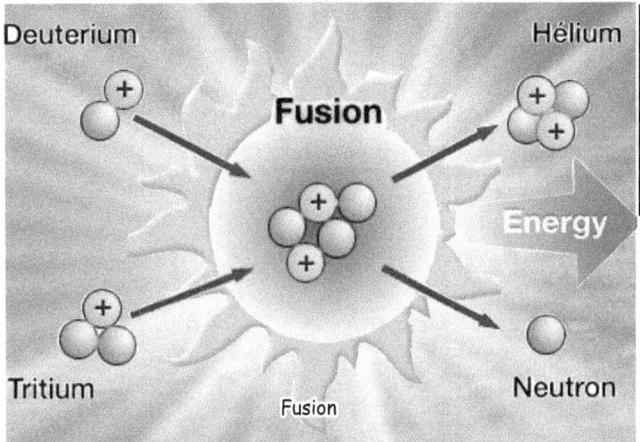

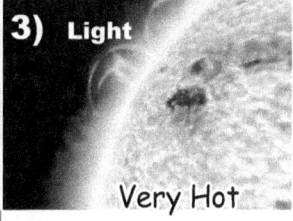

Inside the Sun, with tremendous heat and pressure, two hydrogen atoms join to make one helium atom. This is called "fusion." Heat and light are given off. We see this as sunshine.

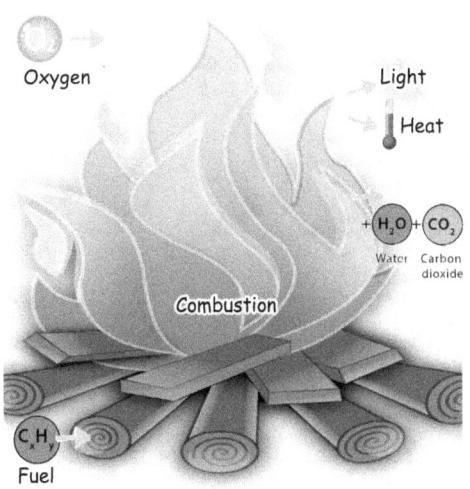

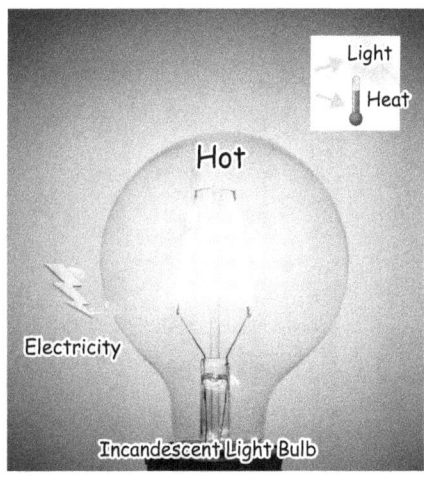

Fire called "combustion" is the input of fuel and oxygen, a chemical reaction and the output of heat and light. Also, water and carbon dioxide (CO2) are given off.

Electricity heats up a thin twisted wire filament until it glows. Lots of heat is given off, so incandescent light bulbs are inefficient.

LEDs and Life Lights

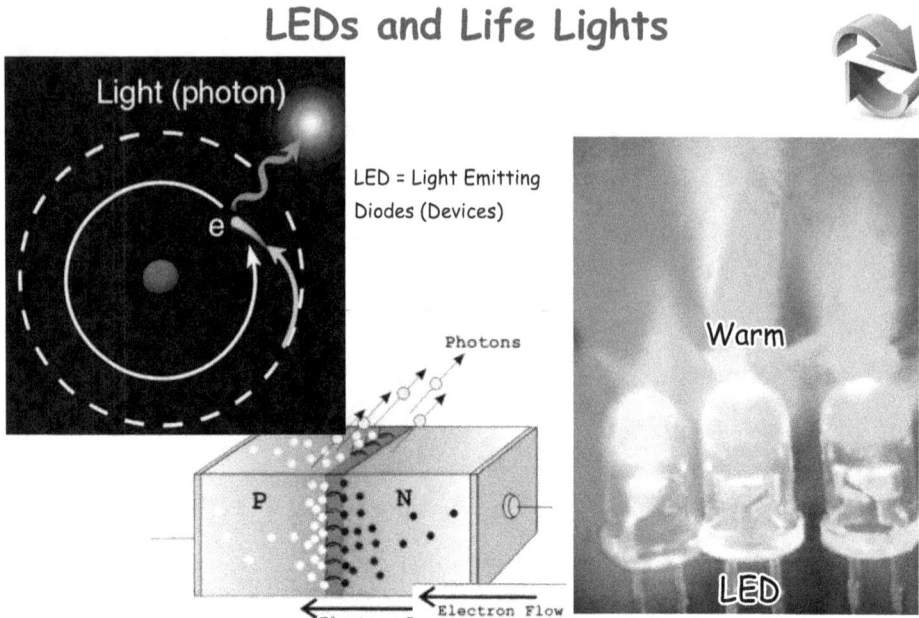

LED = Light Emitting Diodes (Devices)

To recap, electrons orbit around the nucleus center. In LEDs, electrons go from a higher energy orbit to a lower orbit. They emit light photons.

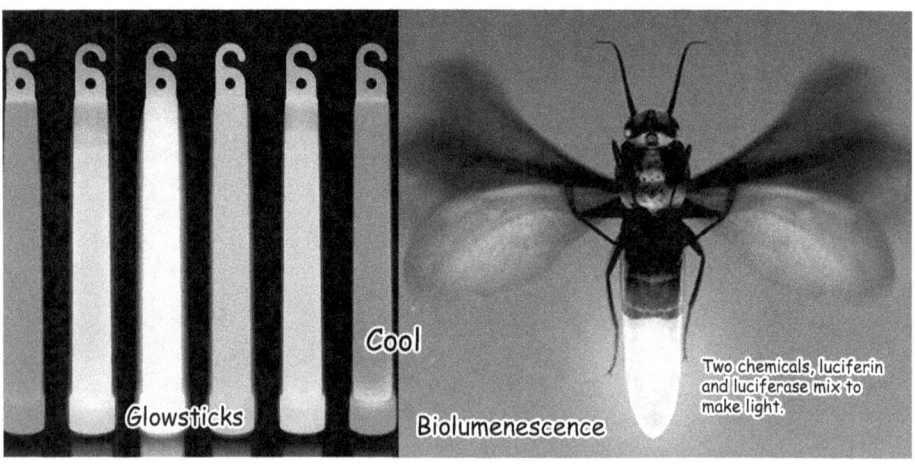

Nature learns that two chemicals (luciferin and luciferase) can mix together and the electrons do the same thing as in LEDs. That is, they go from higher to lower energy orbits and emit light. This is how glow sticks work too.

Teacher - Seven Ideas

4) Life

Wow! DNA directions assemble lifeless atoms into living cells. Every human has trillions of cells that integrate and cooperate altogether. Life is matter with energy in motion.

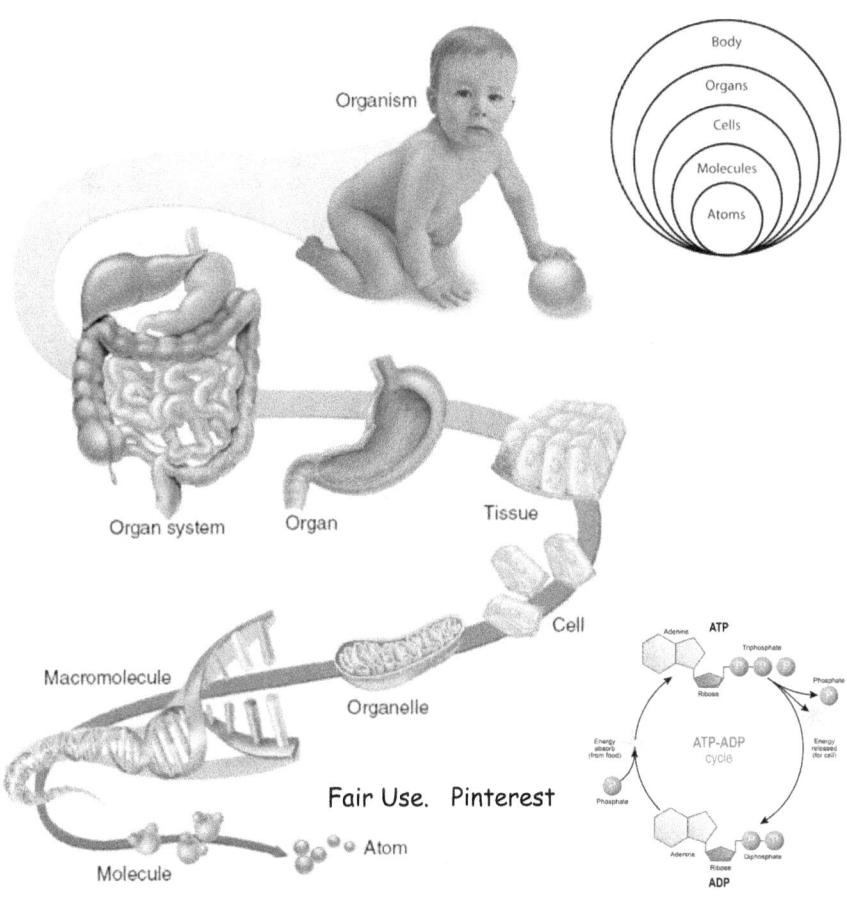

Fair Use. Pinterest

ATP is our energy!

All known life forms get their energy from ATP molecules. It has three phosphate (P) atoms. When a P atom splits off from the molecule then energy is released. This is the energy of life! The ADP molecule with two P atoms is recycled. Energy from our food changes ADP back into ATP that we use as energy and the cycle repeats.

5) Energy

Energy changes forms. We see this when fuel burns and cars drive or airplanes fly. Electricity, magnets, light and sound energies change back and forth into each other to enable our digital devices.

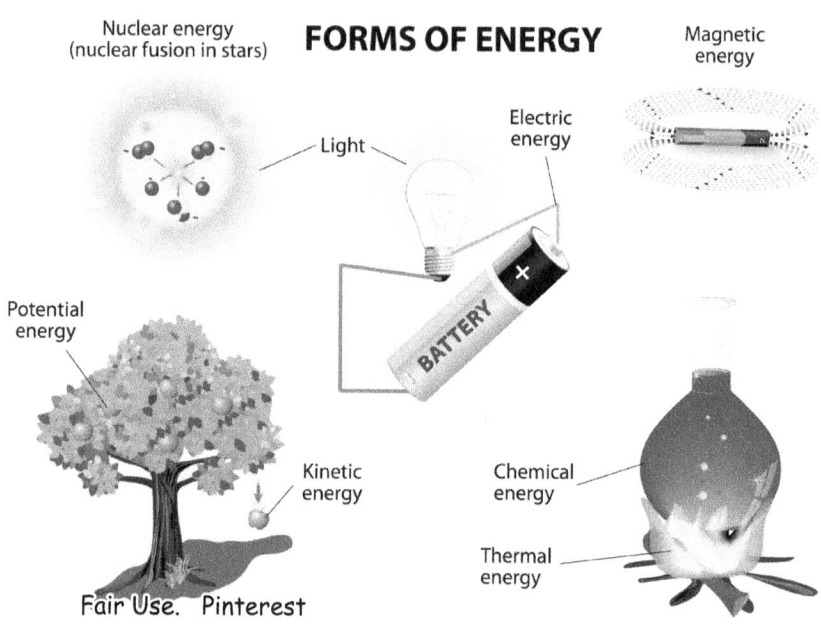

FORMS OF ENERGY

Nuclear energy
(nuclear fusion in stars)

Magnetic energy

Light

Electric energy

Potential energy

Kinetic energy

Chemical energy

Thermal energy

BATTERY +

Fair Use. Pinterest

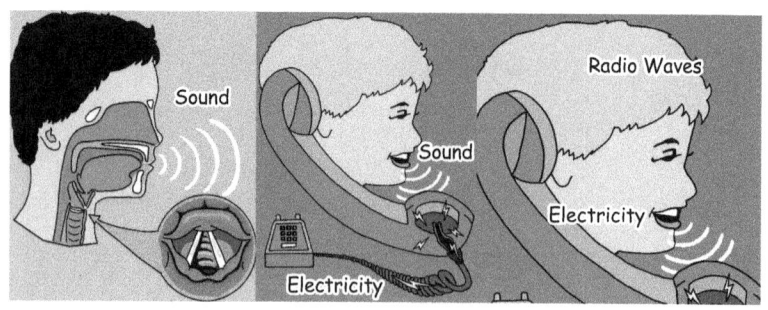

Sound

Sound

Radio Waves

Electricity

Electricity

6) Radio Waves

To recap, electricity flows as currents that make right angle magnetic fields.

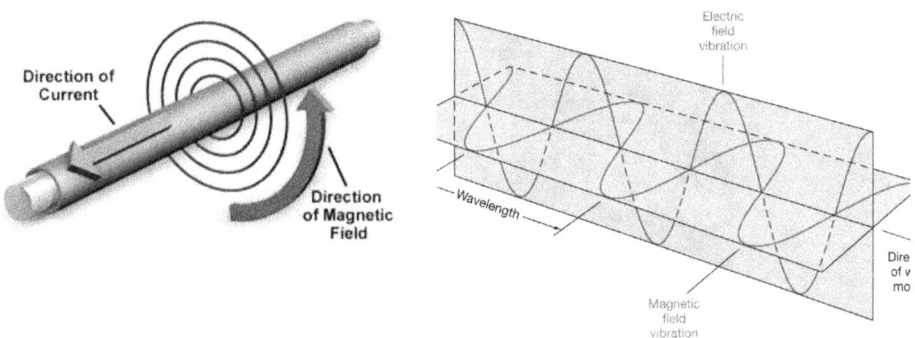

Changing (oscillating) electric (E) currents create changing magnetic (M) fields.

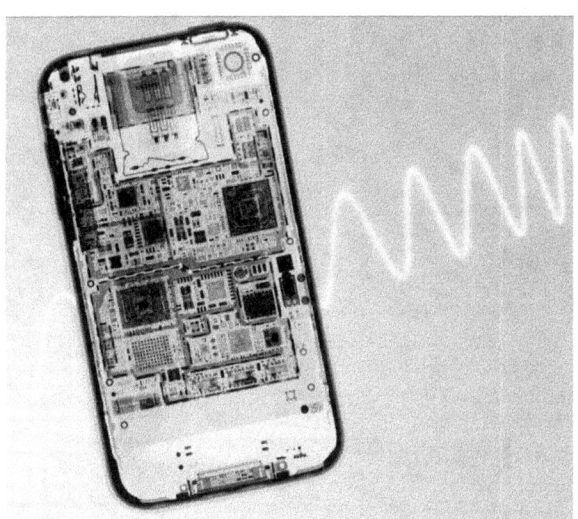

At a smartphone antenna, changing E and M fields create radio waves. That is, R-waves contain electric and magnetic parts that we use to communicate.

7) Gravity

Gravity is big objects pulling on (attracting) smaller objects. For example, the Earth pulls on us and the Moon.

The Sun's gravity pulls on the planets.

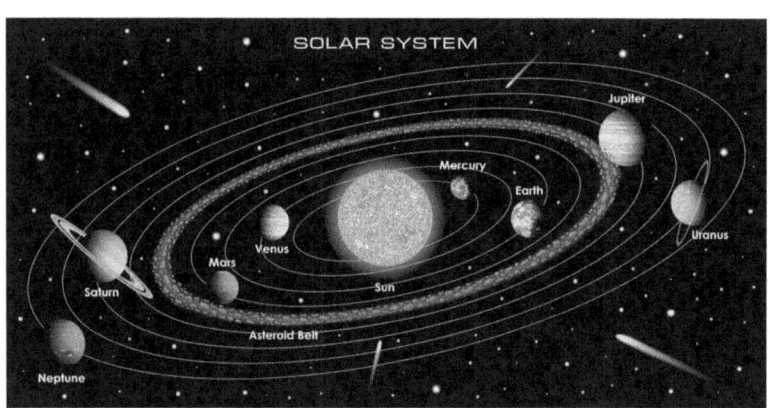

Milky Way Galaxy gravity pulls on our solar system.

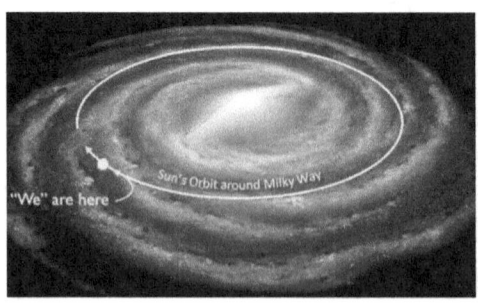

Teacher - Seven Ideas

D) Teacher <u>Reviews</u> Video

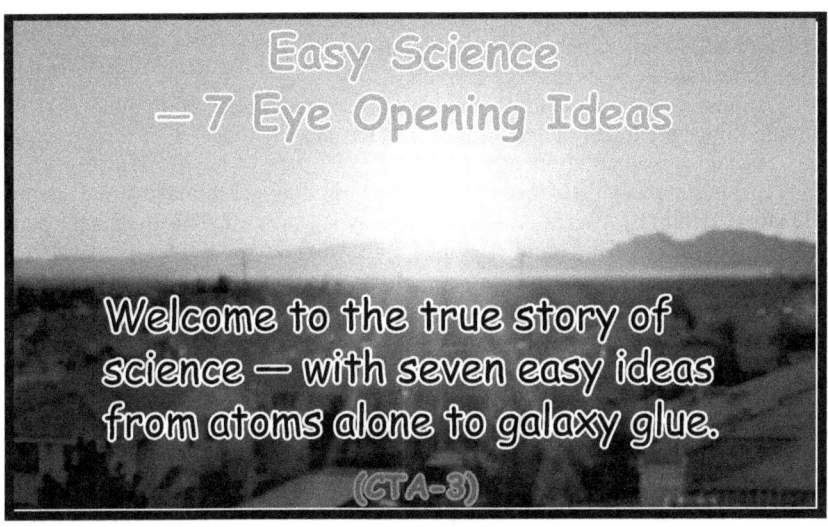

Easy Science
— 7 Eye Opening Ideas

Welcome to the true story of
science — with seven easy ideas
from atoms alone to galaxy glue.

(CTA-3)

For a copy of this video contact:

EDUSTORE
AFRICA

Take Action

7 IDEAS VIDEO Script

WELCOME TO — Science with 7 Easy Ideas
i-1) There is science in space... i-2) and in sunshine.
i-4) Science is in everything from dark to the light.
i-5) Welcome to the true story of Science - with Seven Easy Ideas
From Atoms Alone To Galaxy Glue
1-1) ONE, Everything on earth is made of atoms.
1-2) Actually, everything we see is only made of less
than 100 different types of atoms.
1-3) Atoms combine together to make new chemicals called compounds.
Living things are mostly made of these atoms That includes, plants, animals and
us. The sun is made of atoms too.
1-4) Inside stars, small H (Hydrogen) atoms join together to make a new element,
He atoms (Helium). When atoms join, energy is given off. This makes the sun hot
but it is not on fire. This is called, Fusion. So, when atoms JOIN together, Energy
is given off.
1-5) Interestingly, Energy is also given off, when atoms BREAK apart. This is
called Fission. When Uranium atoms break into smaller atoms, Energy is given
off. The Energy from splitting uranium atoms powers atomic bombs and electricity power plants. How is Electricity made?
2-1) TWO, Electricity and Magnets are related
2-2-1) Want to know a Secret, just turning wires near Magnets to make Electricity!
2-2-2) "It's that simple! All the AC Electricity around me is made JUST this Way."
Water...... Steam........ And Wind powers, all turn the wires near magnets to
make the electricity.
2-3-1) The Reverse, makes electric motors move.
2-3-2) Electricity flows in wires, makes a magnet that pushes on other magnets to make
motors turn. Wow!   These all move because of electric motors.
2-4-1) Do you know? The earth is a giant magnet too.
I see this with a compass.
2–4-2) Electricity and magnets also make Electronics work.
2-4-3) Digital means bits of electricity that power computer chips.
2-4-4) Disk Drives work with Magnets.
2-5) Next, Electricity makes light in Light bulbs.
3-1) THREE, LIGHT, Bends, Bounces and Beams
It is hard to explain what light is. It is easier to see what light does.
3-2) Light bends. We see this with prisms. Also, when light
goes from air to water. Lenses bend light to take pictures too.
3-3) Light also bounces. Light reflects in mirrors. And lakes.
Also, we see the moon because it reflects the suns light.
The sun gives off light.
3-4-1) With hot fusion, the sun gives off light or shines.
The sun should shine, for Four Billion years more.
3-4-2) Some light uses cool chemicals to make its own light.
3-5-1) Light is also why we see. 3-5-2) With light, we also see movies.
3-5-3) Some movies are about dinosaurs.

7 IDEAS VIDEO Script

4-1) FOUR, Life Changes. It is Amazing, that dinosaurs and other creatures once lived on earth. 4-2) Over time, life changes or evolves.

4-3-1) Fossils show us glimpses of interesting life forms of the past.
We wonder, at how the world changes over time.

4-3-2) Why is oil called a fossil fuel?

4-4-1) Let`s go back in time to millions of years ago.

4-4-2) As ancient sea life dies, it settles on the ocean floor.
Get covered with sand.

4-5-1) Over time, heat and pressure, turn the carbon into oil & natural gas.

4-5-2) Oil is buried deep underground. How do our ancestor learn about oil?

4-6-1) Overtime, oil seeps through rock cracks up to the surface.
Ancient people learn that oil burns.

4-7-1) Today, people drill holes into the earth to get oil and natural gas.

4-7-2) We use oil and gas as fuel to get energy.

5-1) FIVE, Energy changes Form 5-2-1) Fuel has Carbon atoms from ancient life.

5-2-2) When fuel burns, chemical energy changes into the heat energy.

5-3-1) Fire has three parts.

5-3-2) Fuel C (Carbon) atoms quickly combine with air, O (Oxygen) atoms.

5-3-3) When the atoms join, they give off heat and light.

5-3-4) Fire powers machines. 5-4-1) for example, Controlled bursts of burning gas, push pistons down to make car engines move.

5-4-2) In a jet engine, air comes in, burns & then quickly pushes the plane forward.

5-4-3) Radio Waves help keep planes from crashing into each other.

6-1) SIX, Radio Waves Are Useful!

6-2-1) Radar sends out Radio Waves. Some of the waves reflect off of objects like airplanes. The reflected waves return to the radar station.

6-2-2) Bounced radio waves, on radar screens, help keep planes safely apart.

6-2-3) Planes and ground stations use radios to communicate.

6-3-1) Guess what Radios use? Yup! They use Radio Waves.

6-3-2) Radio Waves can be as long as mountains are high.

6-3-3) Radios don`t need wires. It is why they are called `wireless`.

6-4) WiFi is another wireless communication. It uses Radio Waves to connect computers together. 6-5) Smartphones use Radio Waves to make calls.

6-6) i-Phones are made by Apple. Why do apples fall off trees?

7-1) SEVEN, Gravity is Attractive

7-2-1) Real apples fall to the earth because gravity pulls them down.

7-2-2) Gravity is what keeps our feet on the ground too.

7-2-3) Gravity is pull power. Heavy objects pull on or attract lighter ones.

7-3-1) Gravity is why the moon, circles the earth.

7-3-2) It is also why the planets go around or orbits the sun.

7-4) Our solar system is part of a galaxy that turns together in space, because of gravity. Gravity is the "glue" that keeps it all together.

C-1) To Close. Gravity is Attractive C-2) Radio Waves Are Useful

C-3) Energy changes form C-4) Life Changes. C-5) Light - Bends, Bounces & Beams

C-6) Electricity and Magnets are related C-7) Everything on earth is made of atoms.

C-8) Today, Our world is CHANGING!

C-9) Science helps us make sense of these world changing events..

C-10) Daily, digital objects get more powerful. Also, computer controlled machines and Robots are doing more jobs.

C-11) Knowing science helps me adapt and thrive to the flood of tech changes!

Easy Ideas

Table of Contents

<u>Main Points</u>
1) Everything on Earth is made of **atoms**.
2) **Electricity** and **magnets** are related.
3) **Light** can bend, bounce and beam.
4) **Life** changes.
5) **Energy** changes form.
6) **Radio waves** are useful.
7) **Gravity** is attractive.

Easy Science — With 7 Ideas

1) Everything on Earth is made of <u>atoms</u>.

a) Atoms Alone (elements)

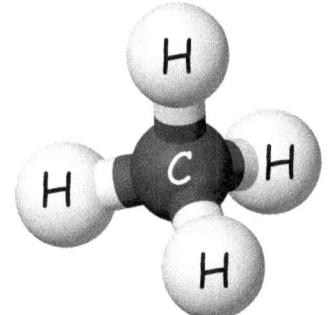

b) Combine into Compounds (Molecules)

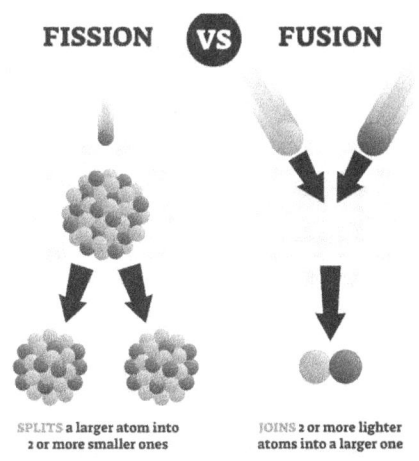

FISSION VS FUSION

SPLITS a larger atom into 2 or more smaller ones

JOINS 2 or more lighter atoms into a larger one

c) Fusion (Fuse) and Fission (Fizz)

1a) Atoms Alone

Human bodies are mostly made of these atoms.

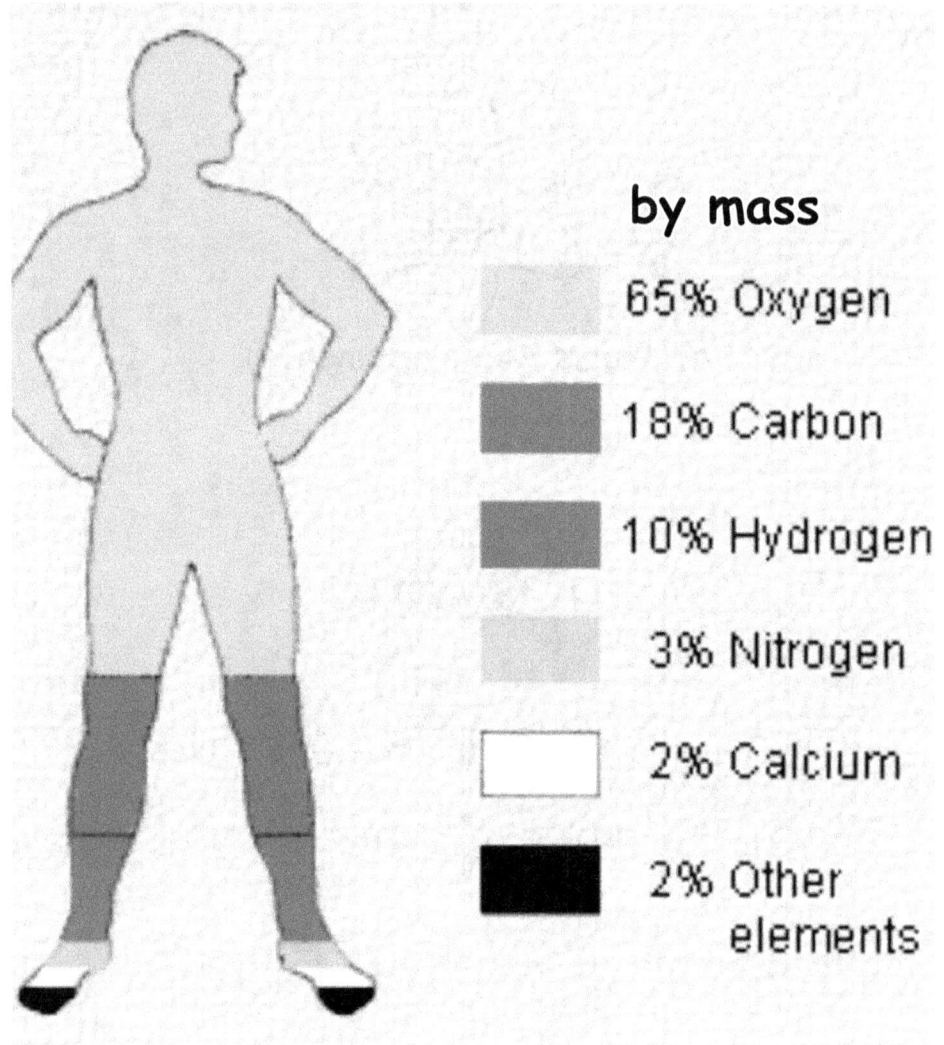

by mass

65% Oxygen

18% Carbon

10% Hydrogen

3% Nitrogen

2% Calcium

2% Other elements

Note that the oxygen and hydrogen join to make water. People are about 75% (3/4th) water.

Teacher - Seven Ideas

— Edible Atoms

Put gumdrops on toothpicks to show
how atoms join to make molecules.
Here are examples CO2, H2O, CH4.

Name of greenhouse gas	Recipe	Shortcut (formula)	Gumdrop model
Nitrous oxide	2 nitrogen atoms and 1 oxygen atom	N_2O	
Carbon dioxide	1 carbon and 2 oxygen atoms	CO_2	
Water vapor	2 hydrogen atoms and 1 oxygen atom	H_2O	
Methane	1 carbon atom and 4 hydrogen atoms	CH_4	

For more info see: https://spaceplace.nasa.gov/gumdrops/en/

1b) Compounds (Molecules)

Atoms join together
to make new chemicals called compounds.

Here are examples of compounds.

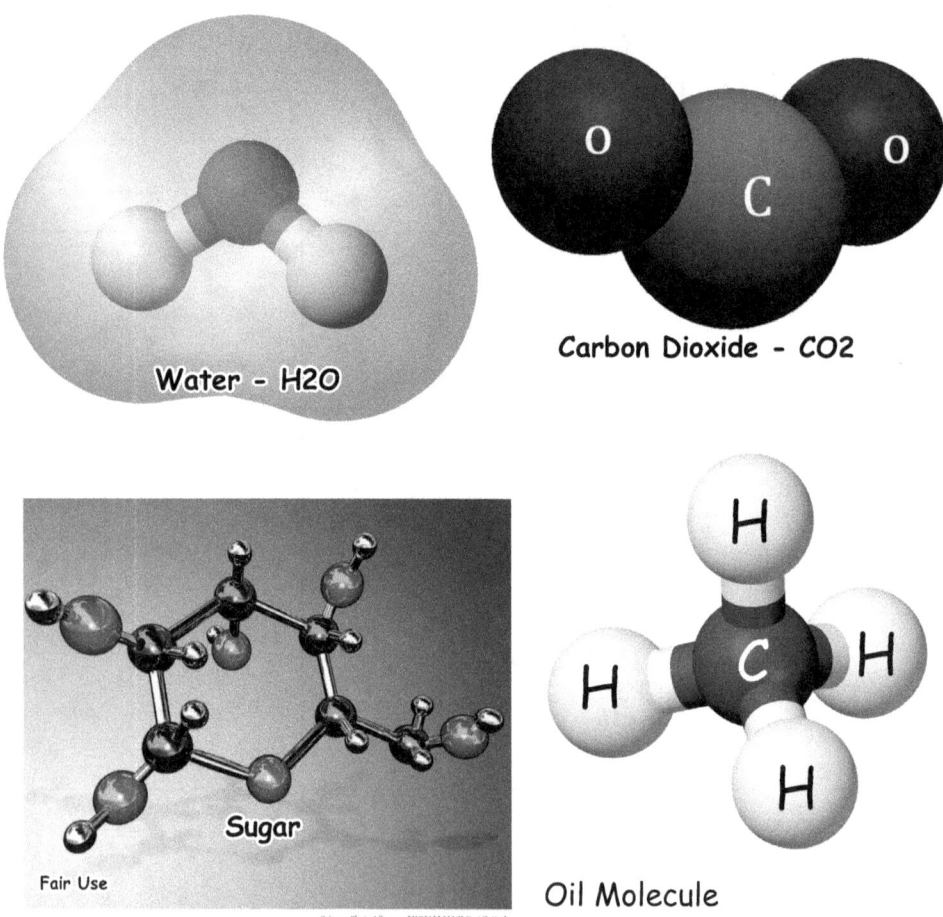

Water - H2O

Carbon Dioxide - CO2

Sugar

Fair Use

Science Photo Library - MIRIAM MASLO. / Getty In

Oil Molecule

Groups of atoms that join together into compounds are called <u>molecules</u>.

Mixes

Atoms and compounds mix together.

We see this when we bake cookies.

— Make Cookies

Start with a list of
ingredients and instructions.

<u>Chocolate Chip Cookies Recipe</u>

2 $\frac{1}{4}$ cups flour (240ml)
1 teaspoon baking soda (5ml)
1 teaspoon salt (5ml)
1 cup soft butter (235ml)
$\frac{3}{4}$ cup brown sugar (180ml)
$\frac{3}{4}$ cup sugar (180ml)
1 teaspoon vanilla (5ml)
About $\frac{1}{2}$ teaspoon water (about 3ml)
2 eggs

A)   Mix together flour, salt and soda.
B)   Stir in butter and both sugars
C)   Add vanilla and water. Mix together
D)   Beat two eggs in separate bowl
E)   Mix eggs into the dough
F)   Stir the dough together until uniform
G)   Add chocolate chips. Can add nuts too.
H)   Bake at 375°F (190°C) for 8 to 11 minutes
      or until brown
I)   Eat and enjoy

1c) Fusion

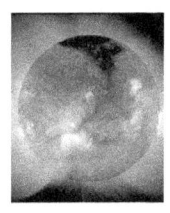

The Sun is hot but it is not on fire. The secret to why the Sun shines is fusion! Inside the Sun, with lots of heat and pressure, light hydrogen (H) atoms join together to make heavier elements of helium (He) and others.

Sunshine is the energy source for Earth life.

One of the meanings of the word "fuse" is to unite or join together.

1c) Fission

Why is nuclear bomb fission
in a book for children's education?

We live in a world full of wonderful,
dream-enabling technology. We also live
in a world with global-scale problems.
Generations to date have ignored the
difficult reality of facing up to and fixing
these problems like the climate crisis and
weapons of mass destruction.
The target audience for this easy science
program will not be able to ignore these
worldwide challenges any longer.
It is the goal of the Science EnLights
Program to instill understanding in young
students, so they will apply enabling
science to fix top global problems.

2) Electro-Magnetism (EM)

Electricity and magnets are related.

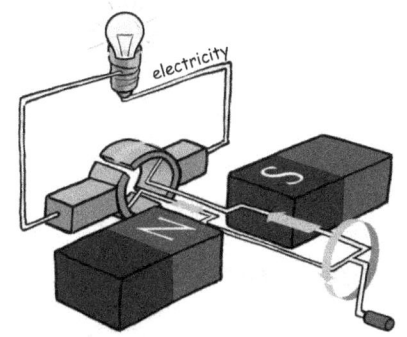

a) Make (Induce) Electricity

b) Electronics

c) Electric Light

2) Battery Electricity

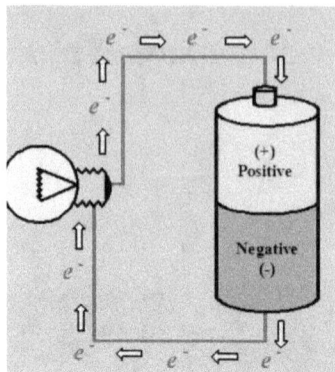

Batteries have electricity.

Batteries power toys, drones and smartphones.

Battery Electricity (DC)

QUESTION?

What are examples of battery-powered objects?

2) Batteries make direct current (DC) electricity.

TRY IT! Battery Electricity (DC)

ANSWER

What are examples of battery-powered objects?

DC

YOU TALK
I
REPEAT

Outlet Electricity (AC)

QUESTION?

What are examples of AC-powered objects?

TRY IT! Outlet Electricity (AC)
ANSWER
What are examples of AC-powered objects?

washing machine

electric kettle

television

microwave

computer

hair dryer

electric fan

vacuum cleaner

refrigerator

2) Outlet electricity powers our homes.

2b) Smartphones use outlet <u>electricity</u> to charge batteries inside that power our electronics.

Electronics

— Lemon Batteries

Put a copper nail (or wire) and a zinc nail into a lemon.
Connect with banana clips to an LED light bulb as shown.

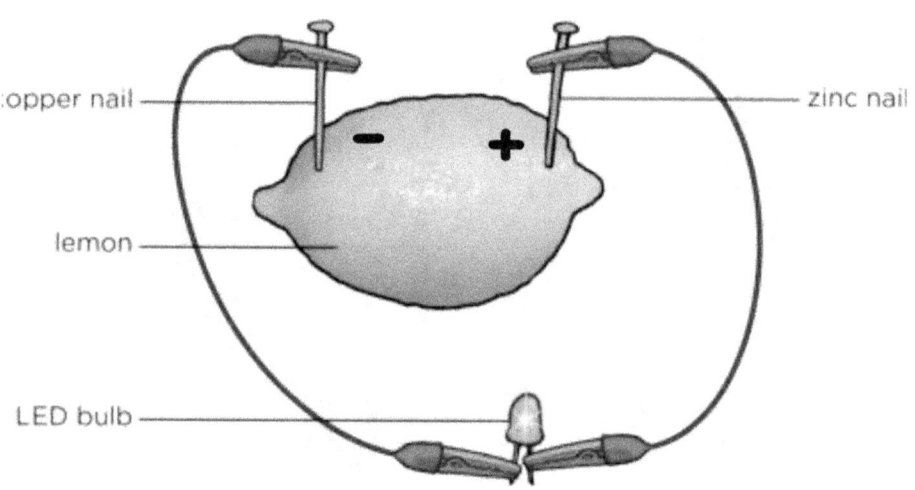

copper nail

zinc nail

lemon

LED bulb

Zinc pulls negatively charged electrons from atoms.
The electrons move through the lemon acid. This makes
the zinc side negative and the copper side positive.
Electricity flows through wires, lights up the bulb
and then flows to the positive side of the battery.

Teacher - Seven Ideas

2) <u>Magnets</u> have two sides called "poles" (north and south).

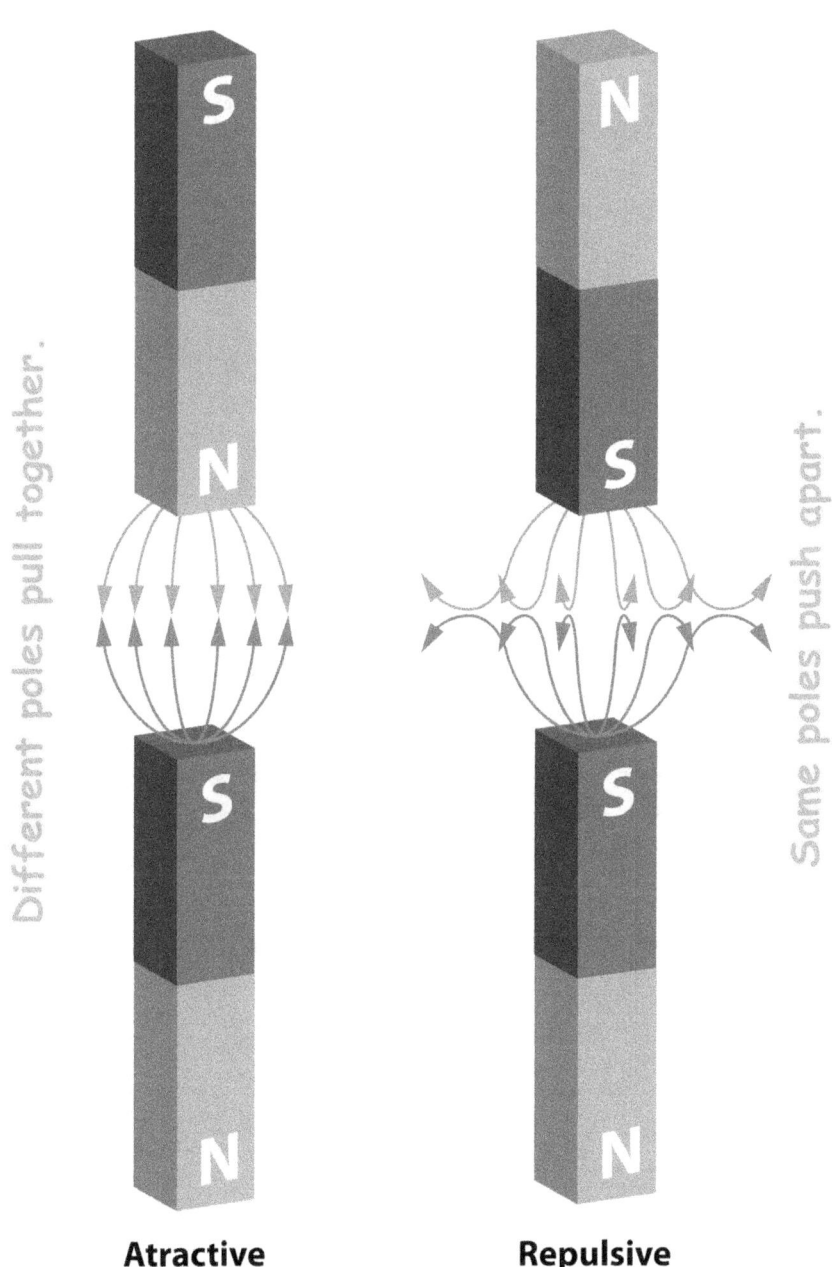

Different poles pull together.

Same poles push apart.

Atractive **Repulsive**

2) Magnets

Magnets have two sides called "poles."

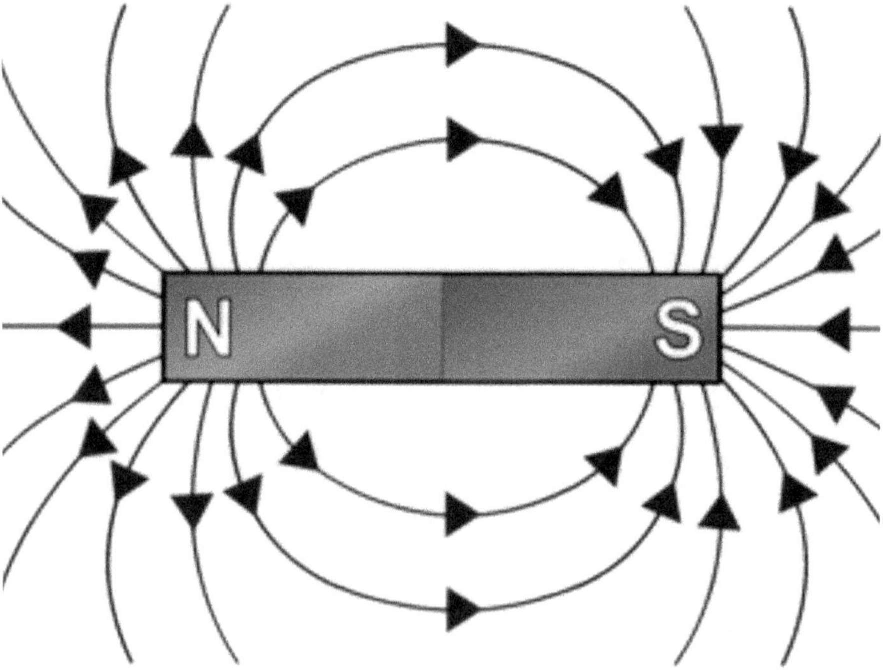

Space around a magnet is called a "field."

— Mag Fields

Purpose: show magnetic fields

Materials: - iron filings
               - cow magnet
               - cooking oil and jar

Steps: 1) Ask, what moves the iron bits?
       2) Pour cooking oil into a little clear glass jar.
       3) Add small amount of iron filings into the oil.
       4) Shake jar.
       5) Hold magnet near th jar.
       6) Explain that the magnet pulls the iron bits
          (filings). We can see the magnetic field.

Related Topics: Earth and compass.
                 North and south magnetic fields.

2) Home Electricity (AC)

Outlet electricity is made by
turning wire coils near magnets.

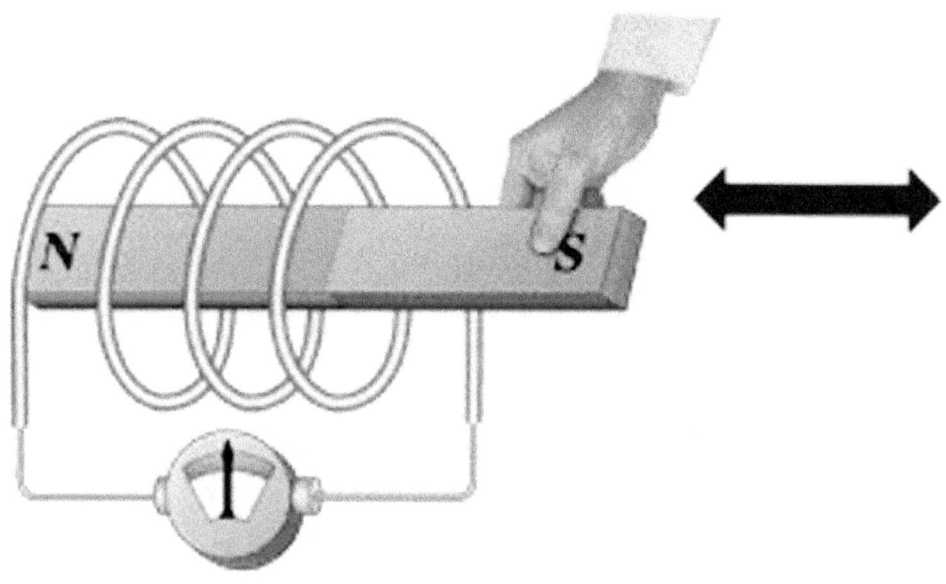

Wow! Electricity around the world is made
by simply turning wires near magnets!

To Recap 2) Electricity and Magnets

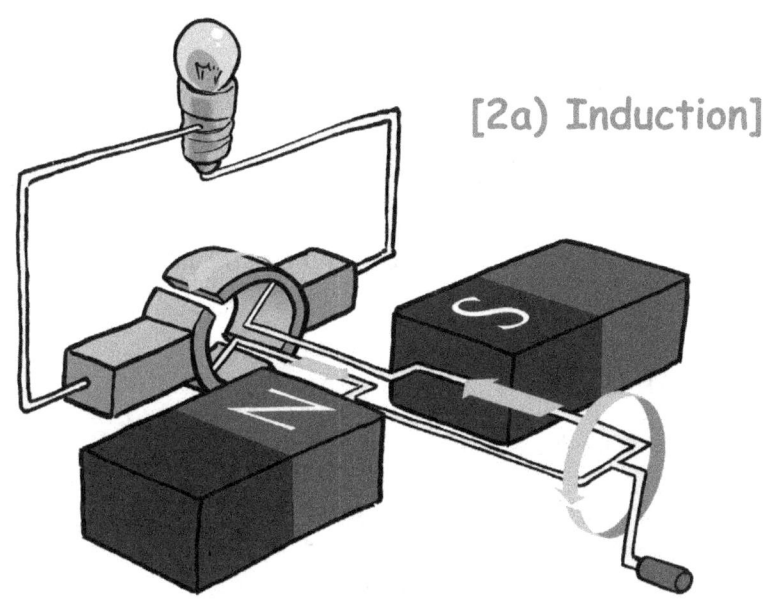

[2a) Induction]

To recap,
electricity is made by moving wires near magnets.

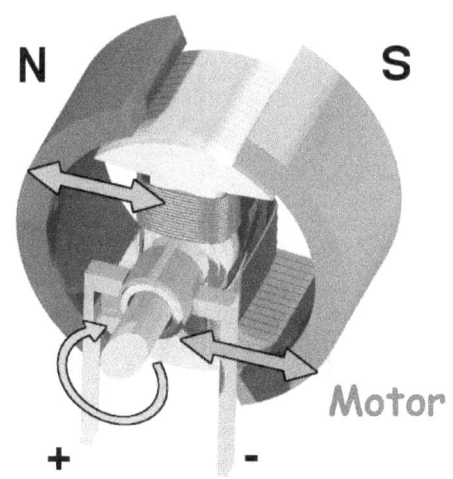

N   S

Motor

+   -

The reverse is true too. Flowing electricity
makes electro-magnets that power motors.

— e-Magnet

Purpose: see how an electromagnet works

Materials: - telephone wire
                    - large nail and paperclips
                    - battery and electricians tape

Steps:
1) How are electricity and magnetism related?
2) Wrap wire around a nail as shown.

3) Cut plastic off of both ends to show copper
4) Use tape to connect one end of wire to battery
5) Make small piles of paperclips
6) Briefly touch the other wire to the battery.
7) Explain when electricity flows, it makes
        perpendicular magnetic fields.

Related Topics: electromagnetic spectrum,
                    radio waves, cell phones

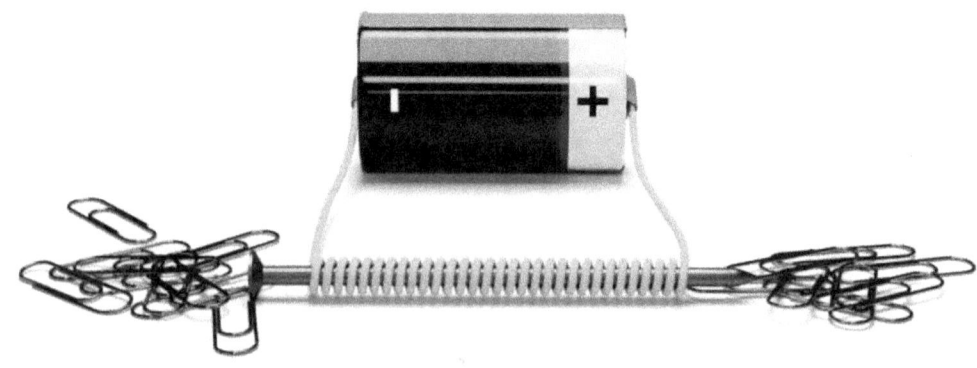

What are examples of electric motors?

TRY IT! Electric Motors

ANSWER

What are examples of electric motors?

For advanced students, identify if AC or DC motors.

3) Light Bends, Bounces and Beams.

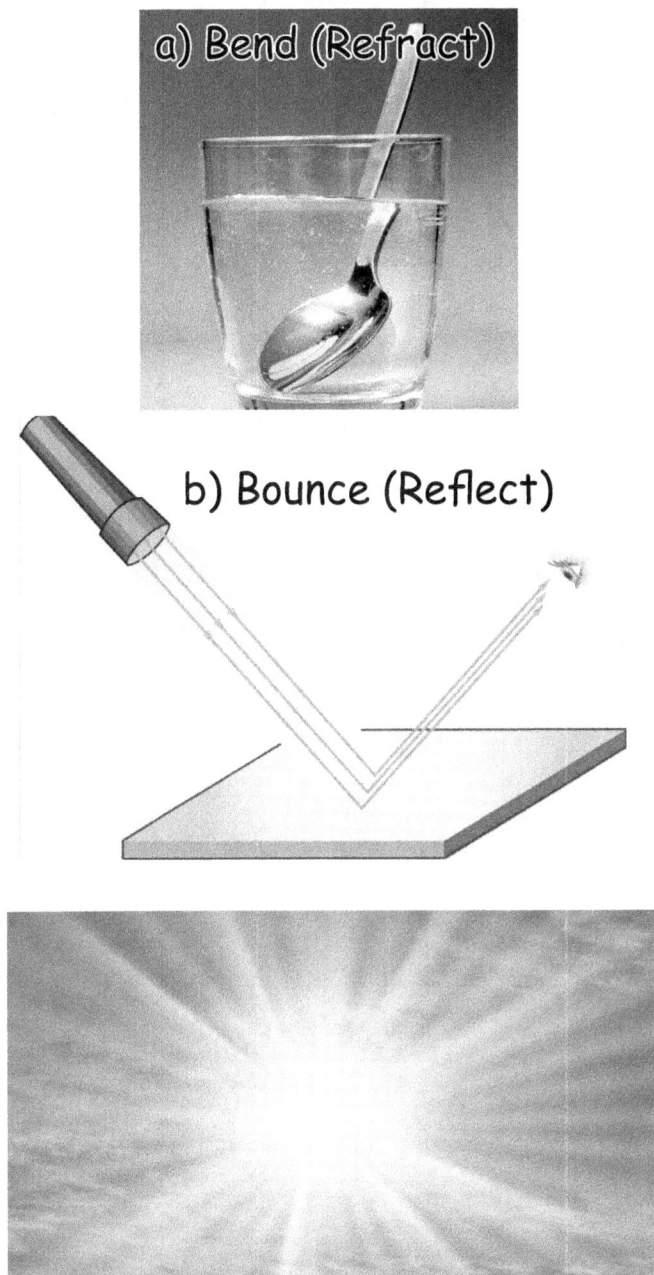

a) Bend (Refract)

b) Bounce (Reflect)

c) Beam (Shine)

Light

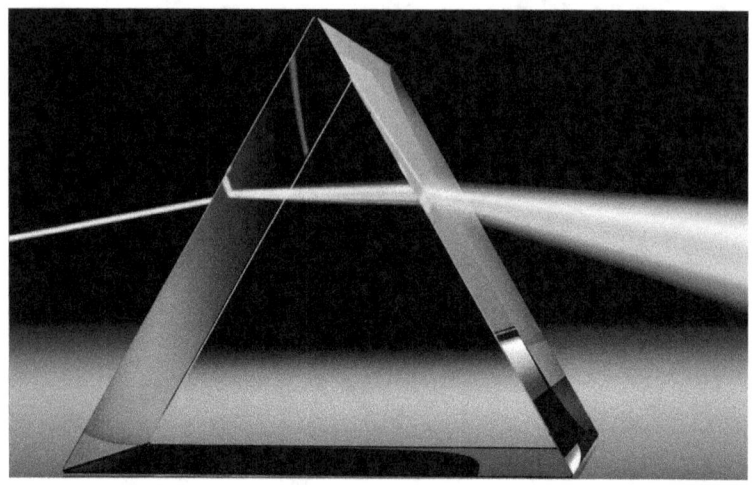

Light bends or refracts through
a prism. The white light separates
into the different colors of a rainbow.

3) Light Makers

What makes light?

TRY IT! 3) Light Makers

ANSWER

What makes light?

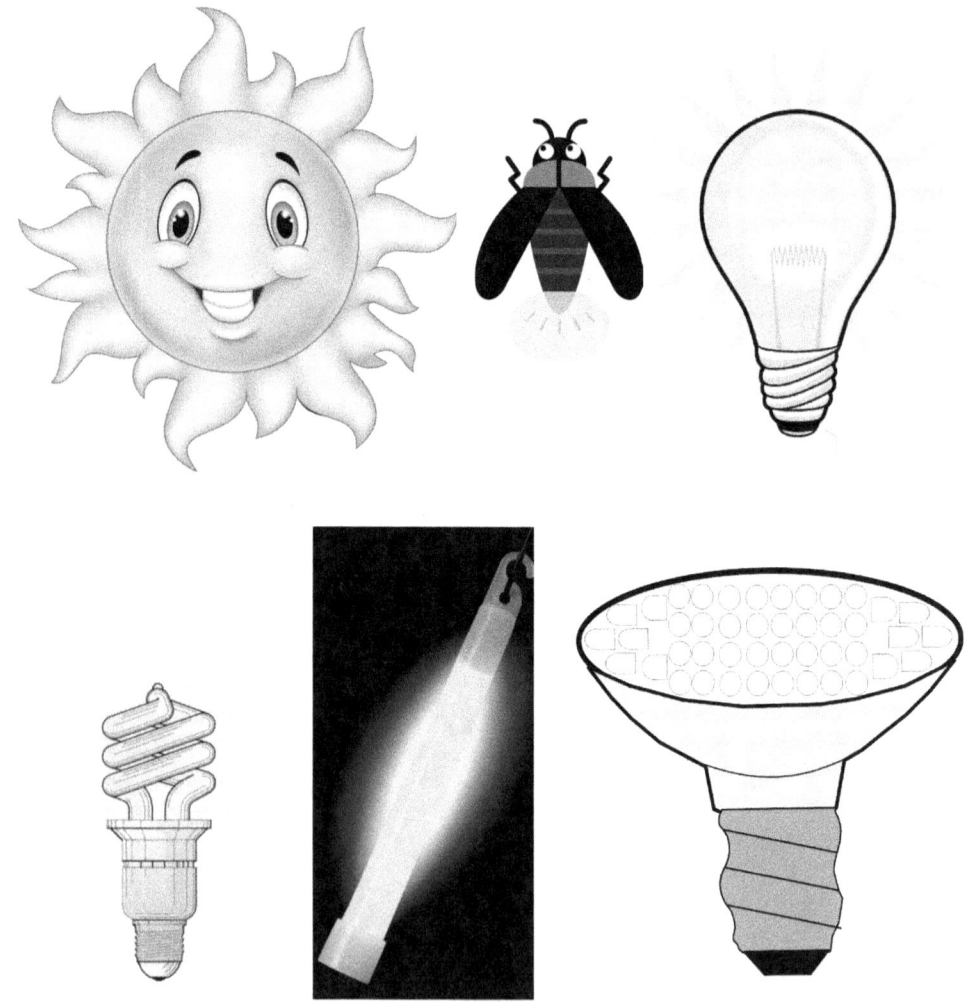

For advanced students, identify how hot light makers are.

3) Light Uses

What do we do with light?

TRY IT!   3) Light Uses

ANSWER

What do we do with light?

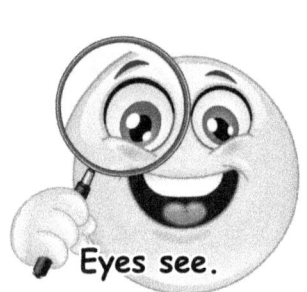

Eyes see.

Plants grow with sunlight.

Screens glow.

Selfies use light.
Cameras focus light
to take pictures.

3) Light Beams and Bounces.

Sunbeams are light.

Light bounces off mirrors.

3) Light Bends

In a prism, white light bends and breaks into colors.

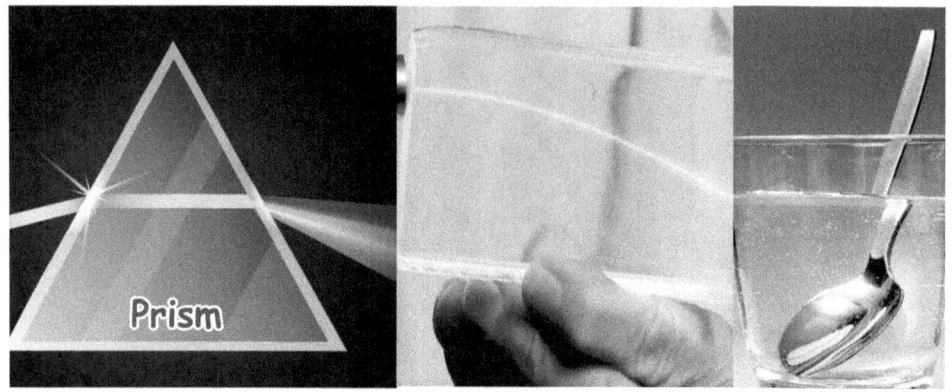

One color laser light bends in this plastic too.
Light bends in water. It makes the one spoon look like this.

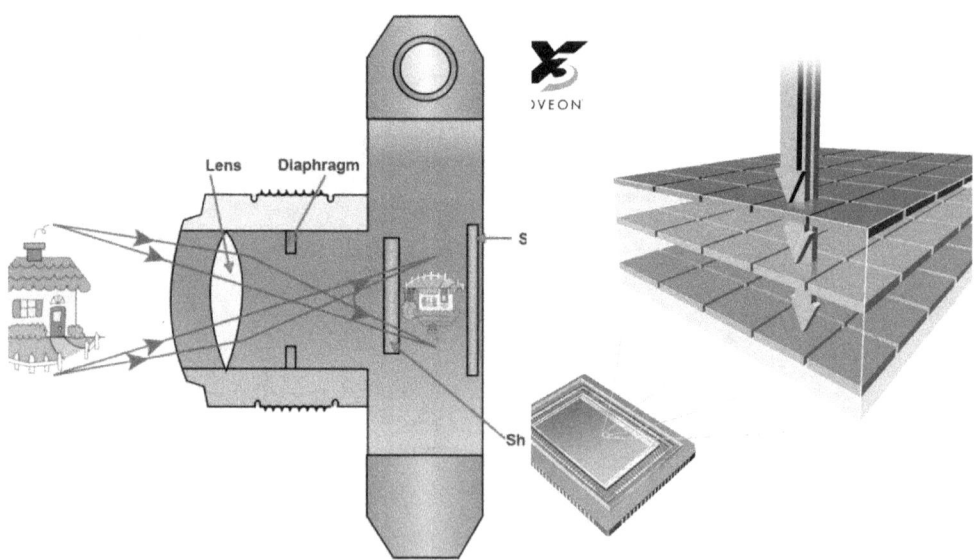

Camera lenses bend light to focus an image. The sensor
changes each pixel micro-dot into red, green and blue parts.

4) Life Changes

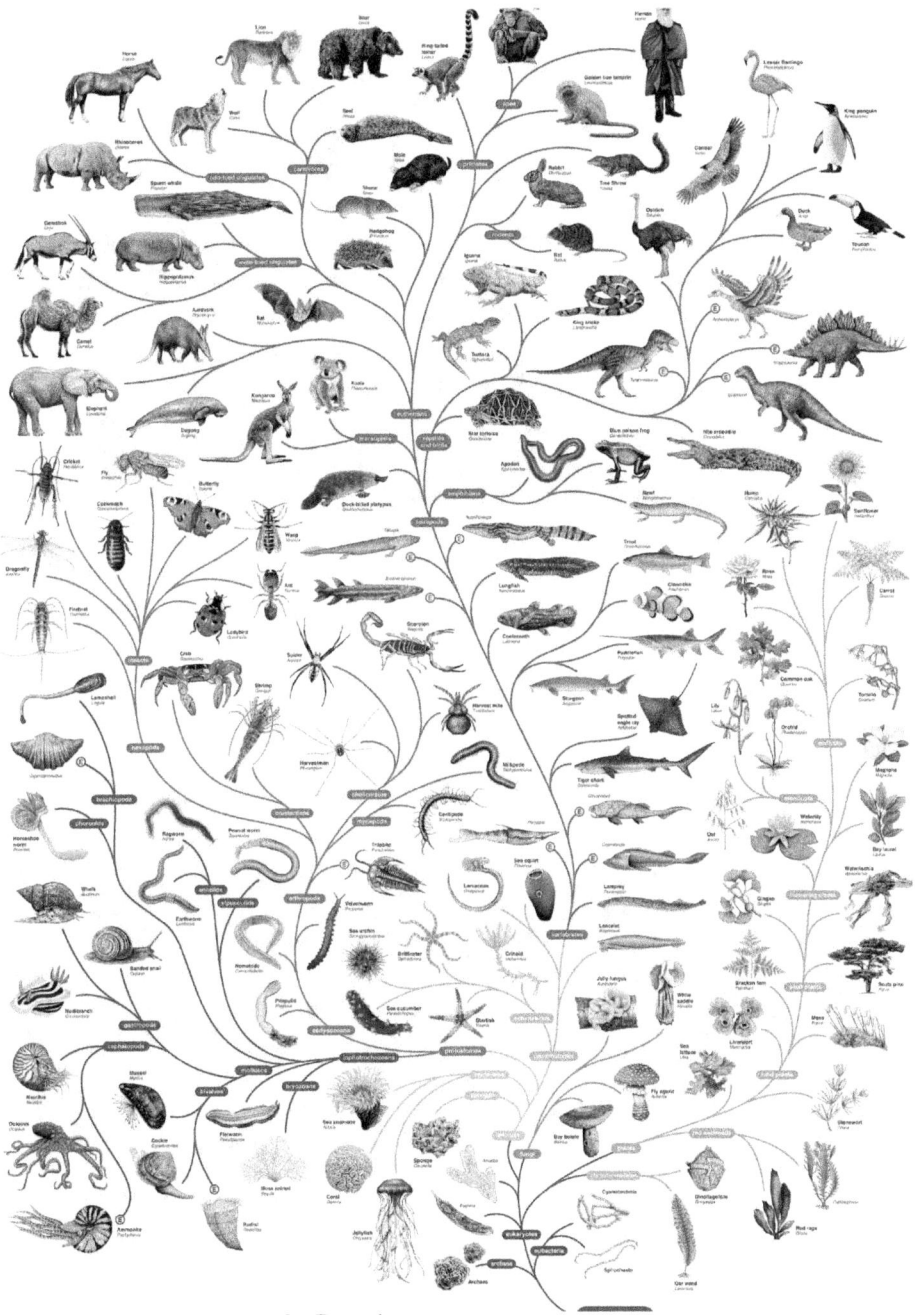

4a) Theory of Evolution states
that Earth life changes over a billion years.

4b) Fossils

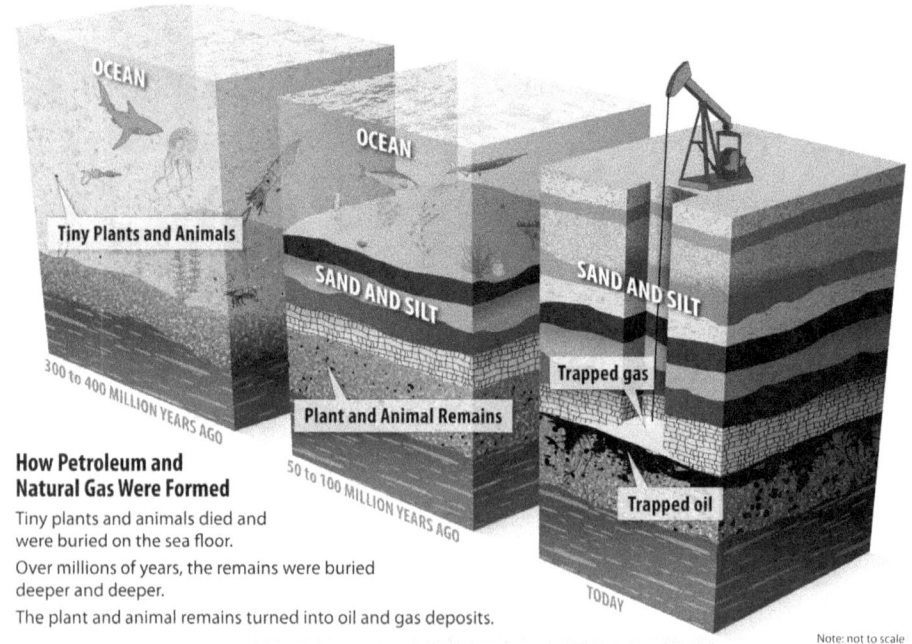

OCEAN

Tiny Plants and Animals

300 to 400 MILLION YEARS AGO

OCEAN

SAND AND SILT

Plant and Animal Remains

50 to 100 MILLION YEARS AGO

SAND AND SILT

Trapped gas

Trapped oil

TODAY

Note: not to scale

How Petroleum and Natural Gas Were Formed

Tiny plants and animals died and were buried on the sea floor.

Over millions of years, the remains were buried deeper and deeper.

The plant and animal remains turned into oil and gas deposits.

4c) Fossil Fuels

We see this when
seeds grow into plants.

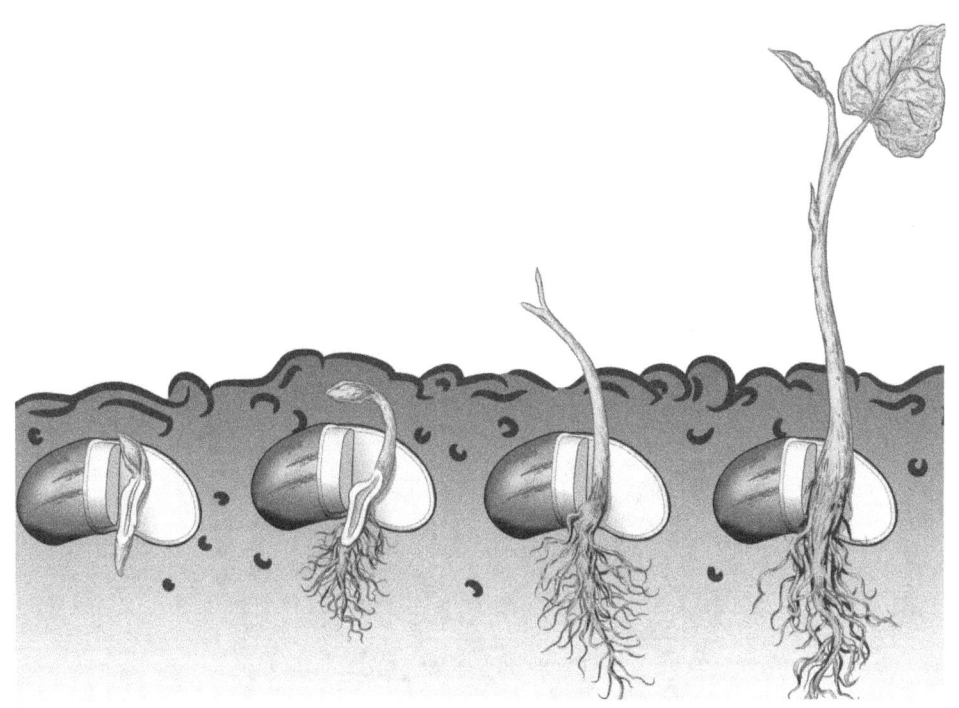

4) Life Changes — Metamorphosis

Here are two more examples.

Tadpoles become frogs.

Larvae become butterflies.

Teacher - Seven Ideas

4) Life

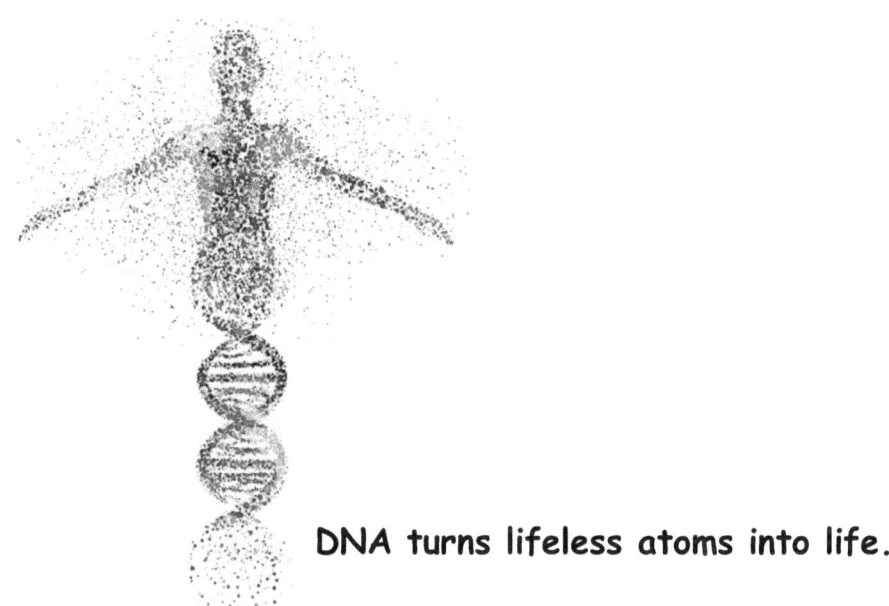

DNA turns lifeless atoms into life.

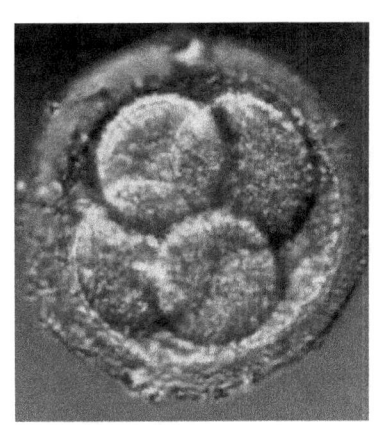

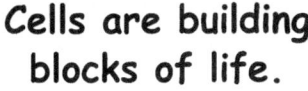
Cells are building
blocks of life.

DNA is like cell software. DNA makes cells. DNA
tells cells how to operate like nano-sized cell apps.

4) Life

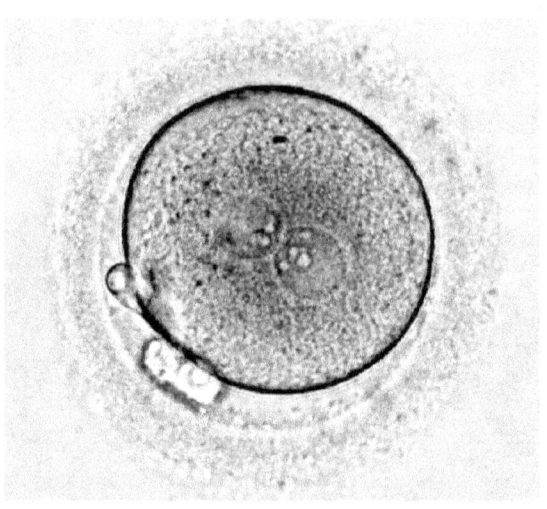

We start as one cell...

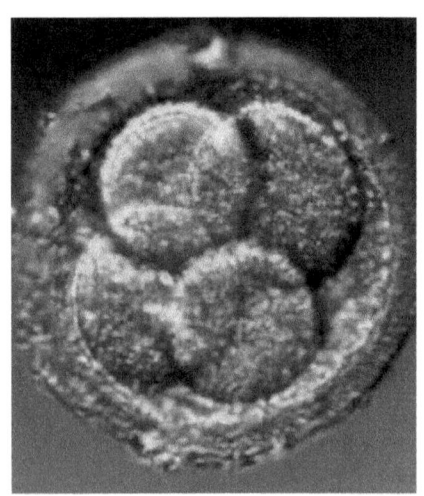 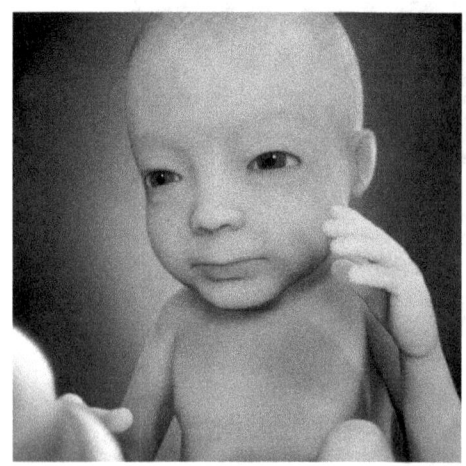

... that turns into trillions of cells.

4) Life

Human babies grow to become adults.

4) Life

To recap, atoms change into molecules
that DNA builds into cells. Groups of cells
become organs that work together to make us.

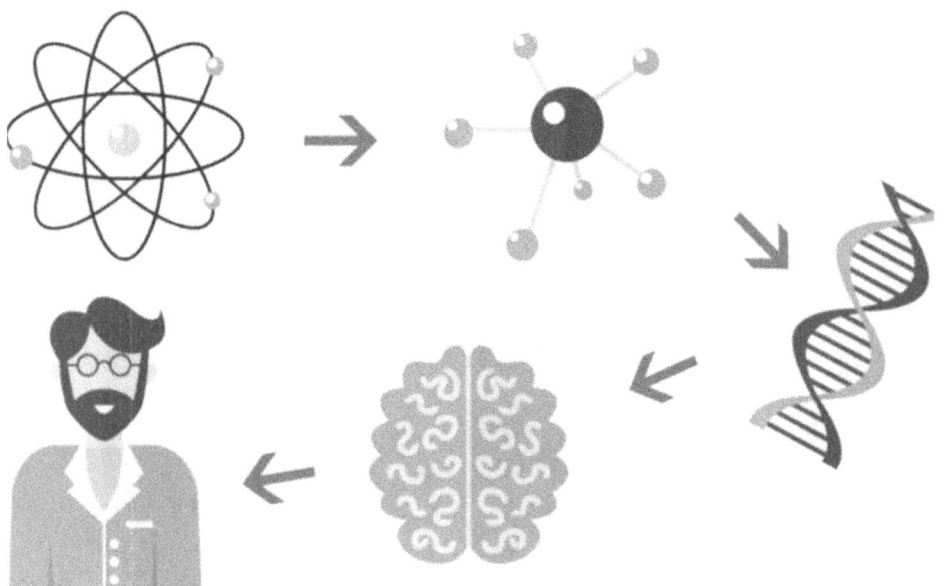

4) Life Change

QUESTION?

How does life change?

TRY IT!   4) Life Changes

ANSWER

How does life change?

examples

Plants grow.

Tadpoles become frogs.

Larvae become butterflies.

Humans change from
babies into adults.

Easy Science — with 7 Ideas

5) Energy Changes Form

a) Fuel Burns

b) Fire Triangle

c) Engines Move

5a) Fuel Burns

Wood has carbon (C) atoms.
Wood burns.

When fuel burns, it changes
chemical energy into the
heat and light energy of fire.

5b) Fire Triangle

Fire has three parts like
a triangle. The carbon (C)
atoms in the fuel quickly combine with oxygen (O)
atoms from the air. When
the atoms join, they give
off heat and light.

5c) Engines Move

Fire gives machines power.

Controlled bursts of burning gas push
pistons down to make car engines move.

5) <u>Energy</u> Changes Gas into Go.

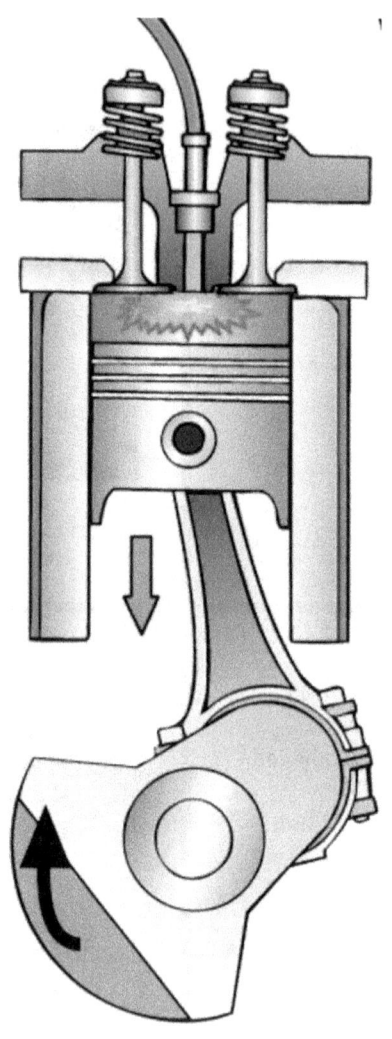

Fuel (gas) burns and engines move.

5) Energy

We see energy with
trains, cars and airplanes.

5) Energy

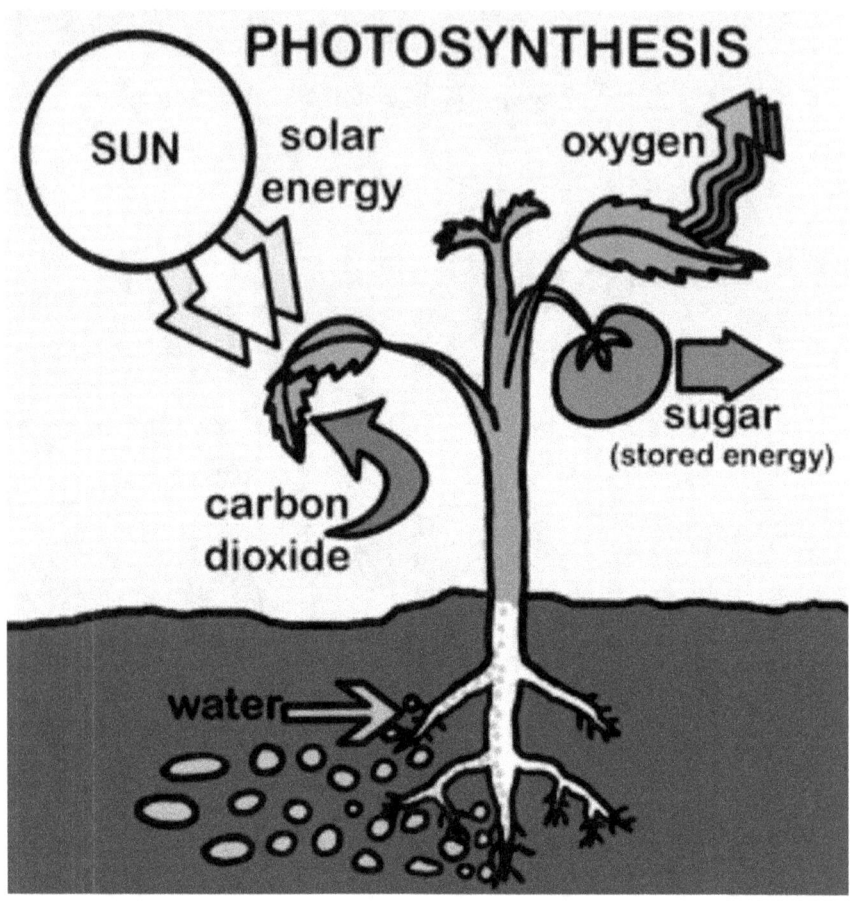

PHOTOSYNTHESIS

SUN
solar energy
oxygen
carbon dioxide
sugar (stored energy)
water

Plants change sunshine into
the chemical energy of food and fuel.

$$6CO_2 + 6H_2O \xrightarrow{\text{Light}} C_6H_{12}O_6 + 6O_2$$

Carbon dioxide   Water   Sugar   Oxygen

This is how atoms flow in photosynthesis.

     Teacher - Seven Ideas

5) Energy Uses

QUESTION?

How do we use energy?

TRY IT!   5) Energy Uses

ANSWER

How do we use energy?

Examples

Plants use sun-
shine to grow.

Fuel and fire power trains, planes and cars.

Rockets too.

People change food into energy.

Easy Science — With 7 Ideas

6) Radio Waves Are Useful.

a) Radar

b) Radio

c) Wireless

Radio Wave Dish

6a) Radar

Radar sends out radio waves. Some of the waves reflect off of objects like airplanes. The reflected waves return to the radar station.

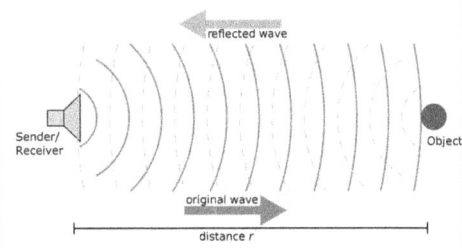
reflected wave
Sender/ Receiver
Object
original wave
distance r

Bounced radio waves turn into blips on the radar screen. They show a planes location and distance. Radar controls airplane traffic.

Planes and ground stations use radios to communicate.

6b) Radio

Guess what radios use? Yup! They use radio waves. Radio waves are "cousins" to light. These waves are called "electromagnetic" or EM. They are made of tiny electric and magnetic parts.

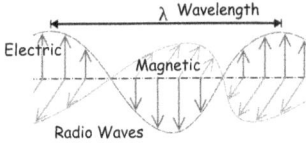
λ Wavelength
Electric
Magnetic
Radio Waves

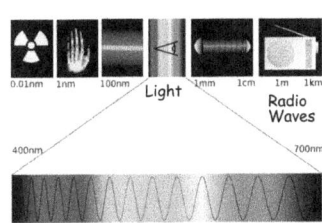
0.01nm 1nm 100nm Light 1mm 1cm 1m 1km
Radio Waves
400nm 700nm

Different types of EM waves vary in length and frequency (number of waves per second).

6c) Wireless

WiFi is a type of wireless communication. It uses radio waves to connect computers together.

Smartphones also use wireless technology.

6) Smartphones use
   <u>radio waves</u> to make calls.

QUESTION?

6) Radio Waves

What everyday objects use radio waves?

ANSWERS

6) Radio Waves

What everyday objects use radio waves?

Examples include: a car radio, remote-controlled car, cellphone, WiFi, Bluetooth.

Teacher - Seven Ideas

7) Gravity

Big objects attract smaller ones.

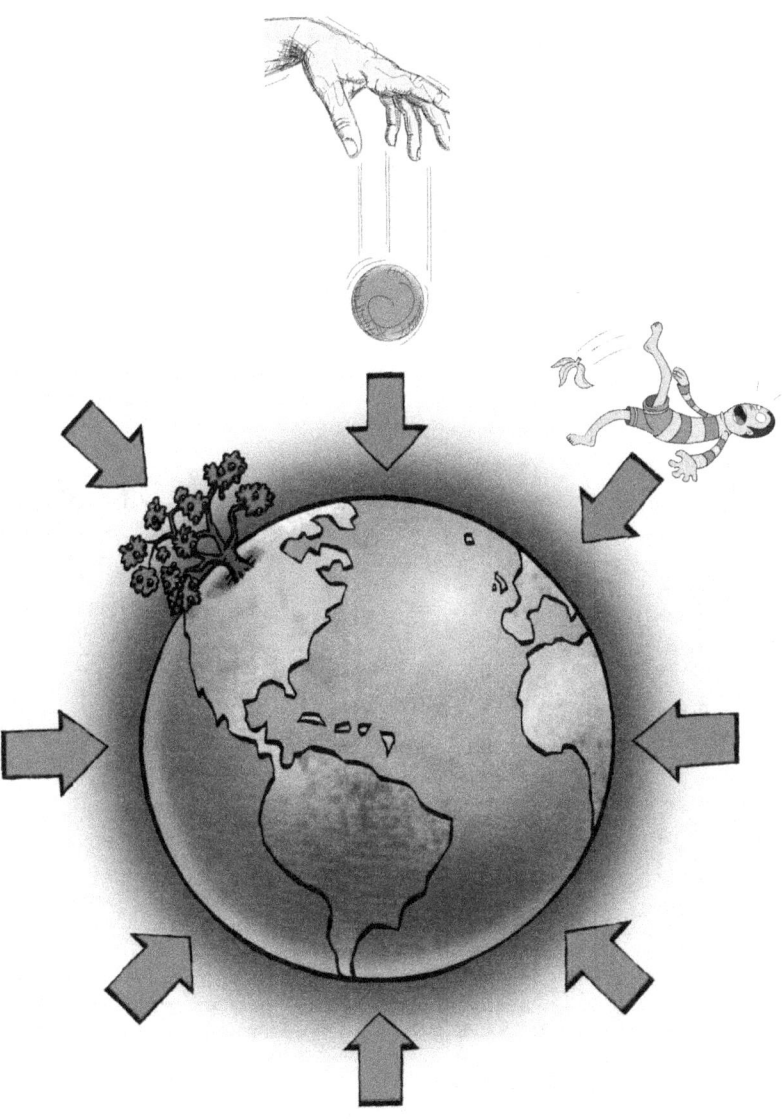

The Earth attracts things like balls, apples and us.

7) Gravity

The Earth attracts the Moon.

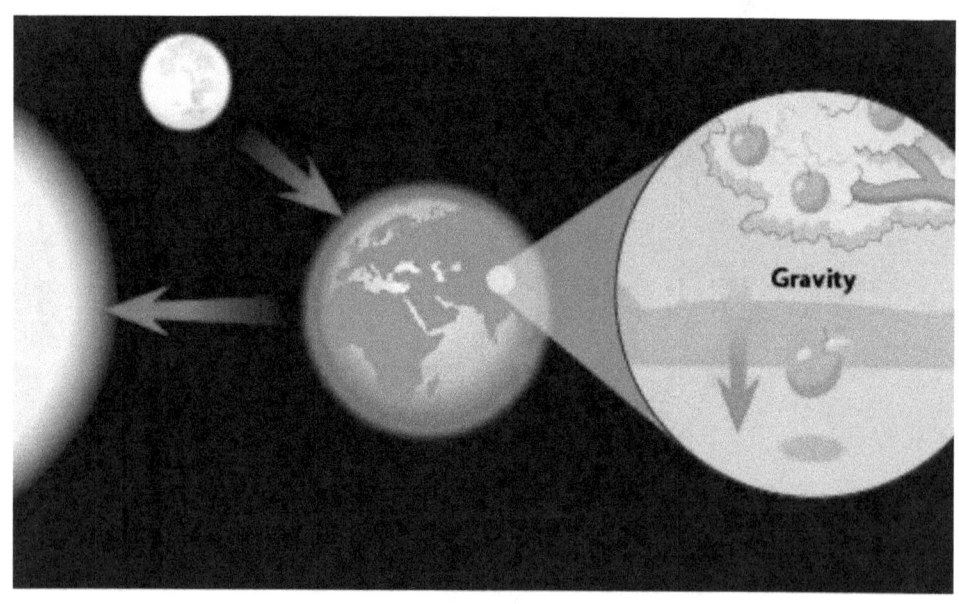

The Sun attracts the Earth and other objects in the solar system.

7) Gravity

If it feels like all this makes your head spin, just keep the following in mind.

The Earth turns on its axis about
1,000 miles (1,600 kilometers) per hour at the Equator.

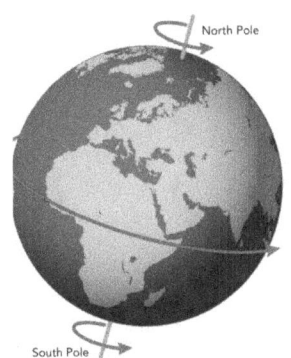

The Earth orbits the Sun at 67,000 miles (107,000 kilometers) an hour.

Our solar system spins around the Milky Way Galaxy center
at about a half million miles (800,000 km) an hour. Every
billion years the Earth circles the galaxy center 4 times.

F) Recap
Easy Science — With 7 Ideas Summary

1) There are <u>atoms:</u> in the Earth,
in our food and in us too.

2) <u>Electricity</u> flows and makes magnets.
Wires turn near <u>magnets</u> to make electricity.

3) Without sun<u>light</u>, space is dark.

4) There are millions of different kinds
of <u>life</u> on Earth that are constantly changing.

5) We use different types of <u>energy</u> to
power machines like electric motors,
gas cars and jet fuel airplanes.

6) <u>Radio waves</u> connect people around
the world with smartphones and the Internet.

7) <u>Gravity</u> attracts objects from
apples and us to the Moon and Earth.

Moving wires near magnets make the AC electricity that we have in our home outlets. This is called "induction." Electricity flows in wires that turn into electromagnets that make motors move. We use electric motors in our home washers, dryers, fridges and other appliances. On airplanes, engines push planes forwards. Also, some of the engine's power is used to turn wires near magnets to generate electricity. People on planes use the electricity to power lights, move air, cook food and the all-important in-flight entertainment screens.

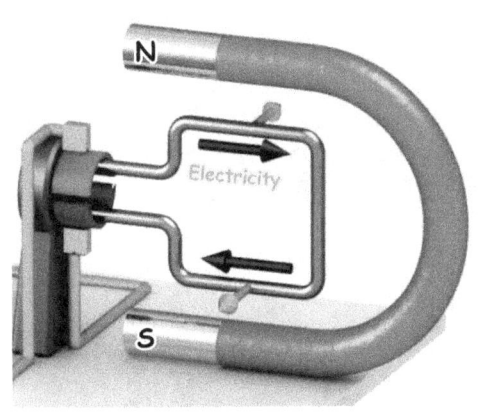

For Advanced Students
— Energy Matters to Us

Objects in motion have kinetic energy (KE). It
is defined by the equation: mass times velocity
(speed) squared. It is why we drive slowly
in a school zone and faster on the freeway.
In Einstein's famous equation, he replaced the
physical speed variable (v) with the speed of
light (c). The equation explains how fusion powers:
the Sun, atomic energy and even the Big Bang.

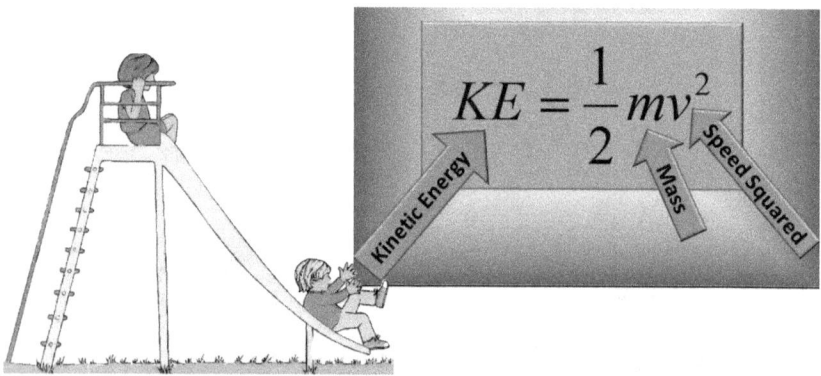

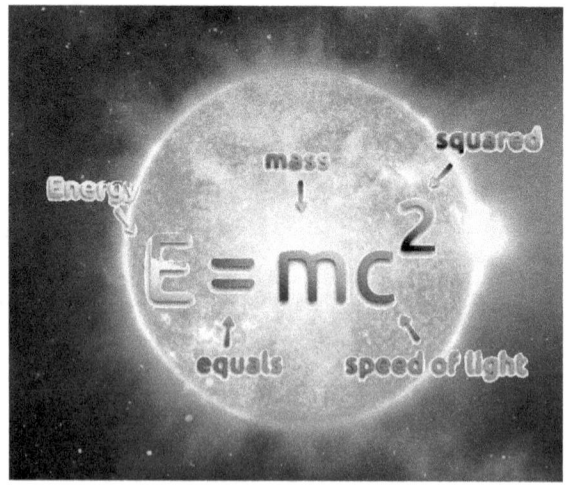

Science Story — Write Stuff

Our everyday objects help us learn and appreciate
science. Let's start with the humble but useful pencil.
In the Middle Ages, sharpened goose feathers
dipped in ink are used to write letters and books.
There is a big storm in Northern England (Lake District).
A shepherd finds a black lump in the roots of a huge tree
that a strong wind blows over. He feels lucky to find what
he thinks is coal. He tries to burn it, but it doesn't burn.
It is a chunk of graphite(carbon).
Millions of years ago, a natural event knocks down a bunch
of huge trees that are buried quickly. Without oxygen
and with lots of pressure, the trees turn into graphite
that much later the shepherd finds.
Miners mine the graphite. Graphite is used to mold
cannon balls. In 1588, England's navy uses
graphite molded cannons on their small, maneuverable
English ships to defeat the Spanish Armada. Later,
someone gets the idea to put graphite inside a
wood sandwich and sharpen it to make pencils.
To summarize, ancient trees are buried quickly. Over time
they change into graphite. Over more time, due to Earth
churns and erosion the graphite logs get back up to the
surface. The graphite is used to make cannon molds that
help keep England speaking English. Graphite cores
surrounded by wooden shells make sharpened pencils
that creative people use to learn things like science.

Integrate

To integrate is to bring together, to make whole. Science integrates energy, atoms, us and the entire multi-verse. Indeed, science explains how energy becomes **atoms** (matter). The energy of **electricity and magnets (E/M)** are related. It is the basis of why batteries power electric motors. Also, why turning wires near magnets make our AC electricity. **Light** is fundamentally photon packets of electro-magnetic energy of different colorful wavelengths. We learn that **energy** can change forms. We see it as chemical energy in food becomes our energy. Also, as electricity becomes **radio** waves that connect smartphones. Sun light enables Earth **life**. Plants capture sun energy to make our food underpinning the cycle of life. Everything from our turning Earth to our spinning galaxy is under the influence of **gravity**. Science helps us connect from the multi-verse to the moving atoms that make up the living cells that enable us.

Teacher - Seven Ideas

Main Points
Science is important to our lives!
1) Air flows in circles for seasons.
2) Water flows from high mountains to low seas and back again.
3) Food energy flows from the eaten to the eaters.
4) Energy flows from fuel to power planes, cars and rockets.
5) Heat flows from hot to cold.
6) Data flows to power our digital devices and cyber Silicon Age.
7) With DNA, lifeless atoms assemble into living cells.

For a copy of this video contact:

7 FLOWS VIDEO Script

i-1) Welcome to Easy Science with 7 Flows

i-2) Every minute, life needs air. There are more atoms in a breath of air then the number of breathes of air on earth.

i-3) Water covers about 2/3rd of the Earth. About 2/3rds of our bodies are water too.

i-4) Food is the fuel of animals and us!

i-5) Applied energy is power from the Sun and machines to ourselves.

i-6) Uneven heat is the source of our seasons

i-7) Data is the electronic heart beats to our digital devices.

i-8) Let's see, 7 flows of Lively Science!

(ONE PAGER without text, Air, Water, Food, Energy, Heat, Data, Life Flows)

1-1)   ONE, Air
Long ago, earth had no oxygen. (Hadean Eon)

1-2)   (Blue-Green Bacteria/Algae) Later, Life evolves that exhales oxygen.

1-3)   Over time, oxygen builds up around the world (text, atmosphere). (Picture $N2 = 78\%$, $O2 = 20\%$)

1-4)   Special microbes turn Nitrogen in the air into soil fertilizer that are plant vitamins (text, Nitrogen Fixation)

1-5)   (Forest) Fire burns oxygen. Animals and us breathe in oxygen. Our cells use oxygen to "burn" food bits into energy.

1-6)   Earth's atmosphere (Text, 100km thick) pushes on us with air pressure
(text, 14.7psi)

1-7)   Sun shines on the earth unevenly (show earth tilt and spin).
Air with different amounts of heat causes our seasons and weather. (Show storms and sunny days)

1-8)   Air moves clouds that are full of water too.

2-1) TWO, Water
The source of Earth's water may have been ancient comet crashes.

2-2) Earth's first life evolves in the oceans.

2-3) The sun powers the water cycle. Sea water evaporates. Clouds move over land. Rain falls and flows down into rivers and back to the sea. This circle (water cycle) waters our planet, plants and people along the way.

7 FLOWS VIDEO Script

2-4) Different sun heat causes different states of water.
(show ice, liquid water and steam)
2-5) Things dissolve in water. This is critical to life
(show plants, animals and people drinking).
Cells need water to function.
We also use water with soap to keep us clean.
2-6) Water can wear away mountains (text, erosion).
Overtime, flowing water made the Grand Canyon.
2-7) With gravity, water flows from high to low places.
We see this in our homes (show tap, use, drain)
2-8) Water helps us prepare our food too.

3-1) THREE, Food
Food is chemistry in action for life.
3-2) Solar powered plants, turn sunshine into sugars (text carbs).
Plants store energy as carbs like fruits and veggies.
3-3) Animals eat the plants. Animals turn carbs into energy.
Animals store energy as fat.
3-4) Omni eaters like people eat both plants (text, carbs) and animals
(text, fats and proteins). We get carbs and vitamins from plants. We
get fats and protein from meat.
3-5) Much of our bodies are made of protein. Here are some examples. (muscles, organs, tissues?)
Cells use proteins (show amino acids) like work orders to do things
like: digest food; (show one -amylase,lipase, pepsin), help us grow
(hormones) and fight germs (show antibodies).
3-6) Humans have about 5 Liters of blood cells.
Every day, blood flows in a circle around our body more than a 1,000
times.
3-7) Nature is full of makers (plants) and takers (animals).
All life is connected in a food web from microbes to me.
3-8) We get our energy by eating food.

4-1) FOUR, Energy
Plants change sunshine into chemical energy.
4-2) Fire changes chemicals in fuel into flame energy.
(Text, combustion)
4-3) Fire and Fuel power: planes; trains and cars.
Rockets too!
4-4) With Gravity, high things have potential energy.
This is why, mountain water wants to flow to the sea.

BONUS-2

Focus on the main point that
our everyday objects have
origins enabled by science.

SCIENCE THINKS!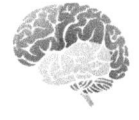

The way that science energizes going from
noticing a problem to thinking about it and
trying solutions is called the "scientific method."

**Science helps us think about our problems
and creatively come up with solutions.**

Let's see how science goes from thinking to seven everyday things: . electricity . light . camera . car . airplane . phone . rocket.

For a copy of this video contact:

1) Easy Ideas
—Teacher Guide—

We start our science quest with 7 fun ideas that open our minds from swirling atoms to gravity pulled apples.

Science

EDUSTORE AFRICA

Indē Ed Project
Charitable Org'n

1)
Science of
Seven Eye – Opening Ideas

FL1 – Sergiu Bacioiu

Douglas J. Alford
Sally Kimangu

Seven Ideas

1) Seven Ideas

Our science quest starts with seven ideas that open our minds — from atoms to apples.

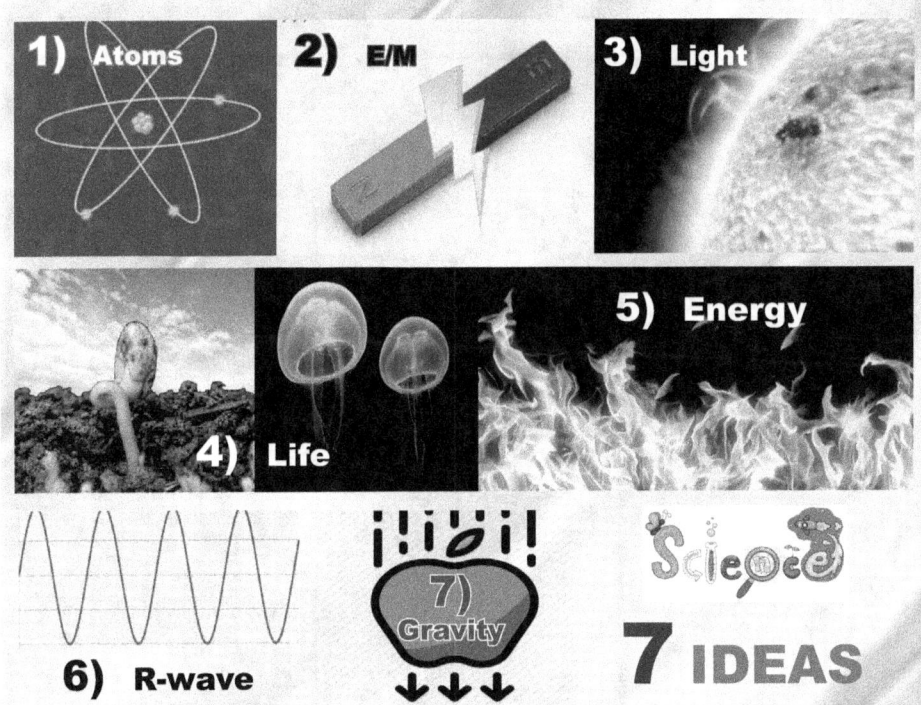

1) Atoms
2) E/M
3) Light
4) Life
5) Energy
6) R-wave
7) Gravity
Science
7 IDEAS

Easy Ideas 1

Airplanes 2

Cars 3

Computers 4

Smartphones 5

Food 6

Nature 7

Space 8

Light 9

AI 10

STEM-Zen Program

© Copyright
Indē Ed Project
Non-Profit, Charitable Org'n
2023. All rights reserved.

Everyday Objects

Bonus

Seven Ideas

Science of
Seven Eye – Opening Ideas

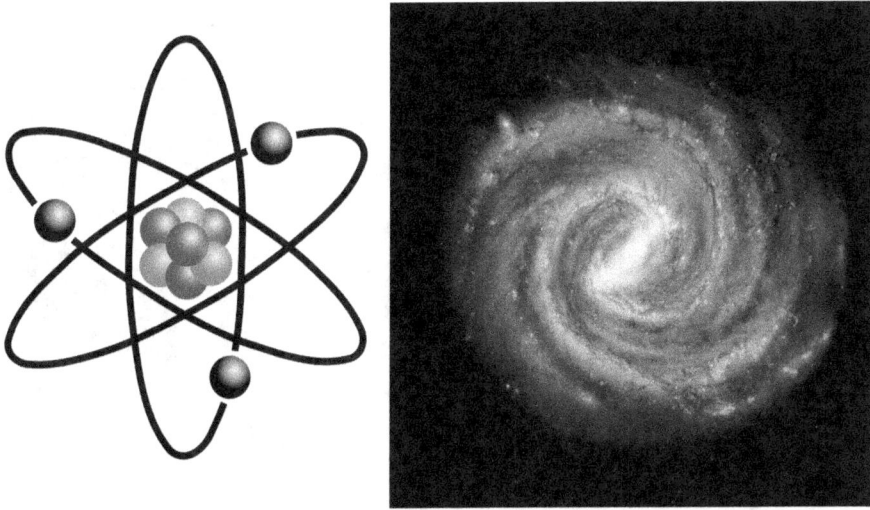

Indeed, science has a dark side when used as a weapon. Atomic bombs are one example, but those same atoms also make electricity. Science is the essence of light too. This is the story of seven scientific ideas that make our world a better place to live.

Science of
Seven Eye — Opening Ideas

Table of Contents

Science of
Seven Eye — Opening Ideas

Table of Contents - continued

To the Readers:
Bits of Science (Sci-Bits) are like
puzzle pieces that fit together.
They help us understand our world.

It is okay if you do not understand every word
in this book. Keep learning Sci-Bits to see the
overall picture of how the world works.

Seven Ideas

Introduction

Some of our ancestors lived in caves.

Seven Ideas

They learned science.
They moved out of the
caves and progressed.

Here are seven ideas of
science that enhance our lives.

Science opens our eyes
and encourages curiosity to
explore our world and beyond.

1) Everything on Earth is Made of Atoms.

a) Atoms Alone

b) Combine into Compounds

c) Fuse and Fizz

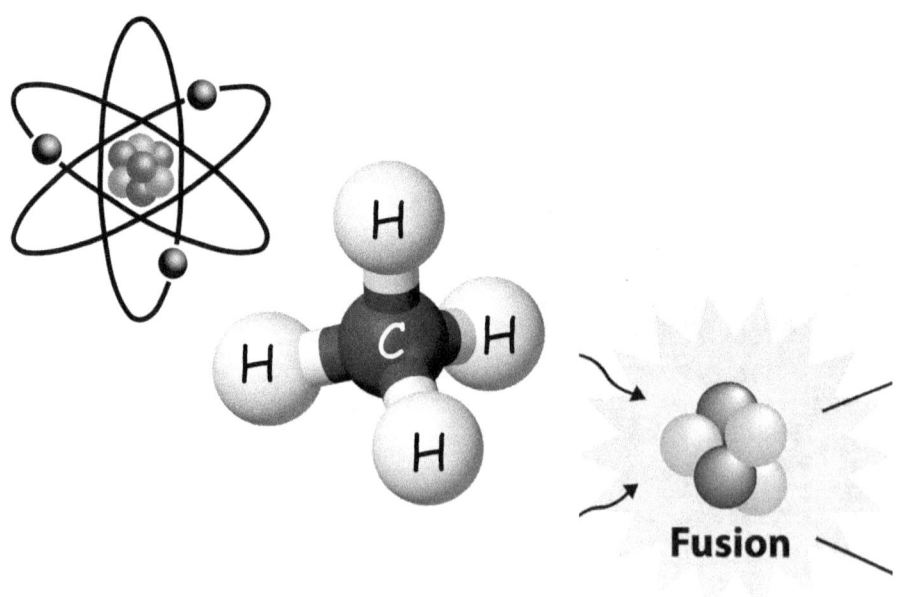

1a) Atoms Alone

Caves and everything else
on Earth are made of atoms.

Caves are made from rocks. Rocks break down to become sand. But what is sand made of? Atoms. It is amazing to realize that everything in the world is made of only about 100 different types of atoms called "elements." Oh, and by the way that includes us too.

Atoms have a central nucleus surrounded by electrons.

1b) Combine into Compounds

Atoms combine together to
make new chemicals. They are
called "compounds" or "molecules."

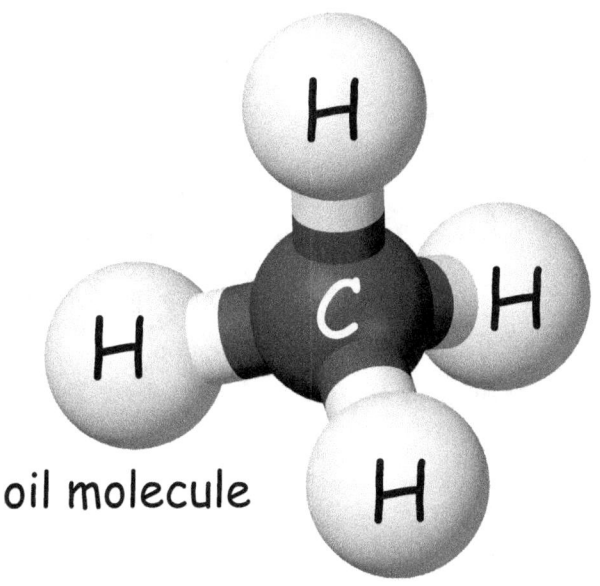

oil molecule

For example, an oil molecule is made of
carbon (C) and hydrogen (H) atoms. The
same atoms under different heat and
pressure make different materials. Carbon &
hydrogen atoms also make sugar, coal and tar.

Mix Together

Atoms and compounds mix together. You see this in chocolate-chip cookies and in hamburgers. Cars & airplanes are mixes too.

Do you know that the Sun is made of atoms too?

1c) Fuse and Fizz

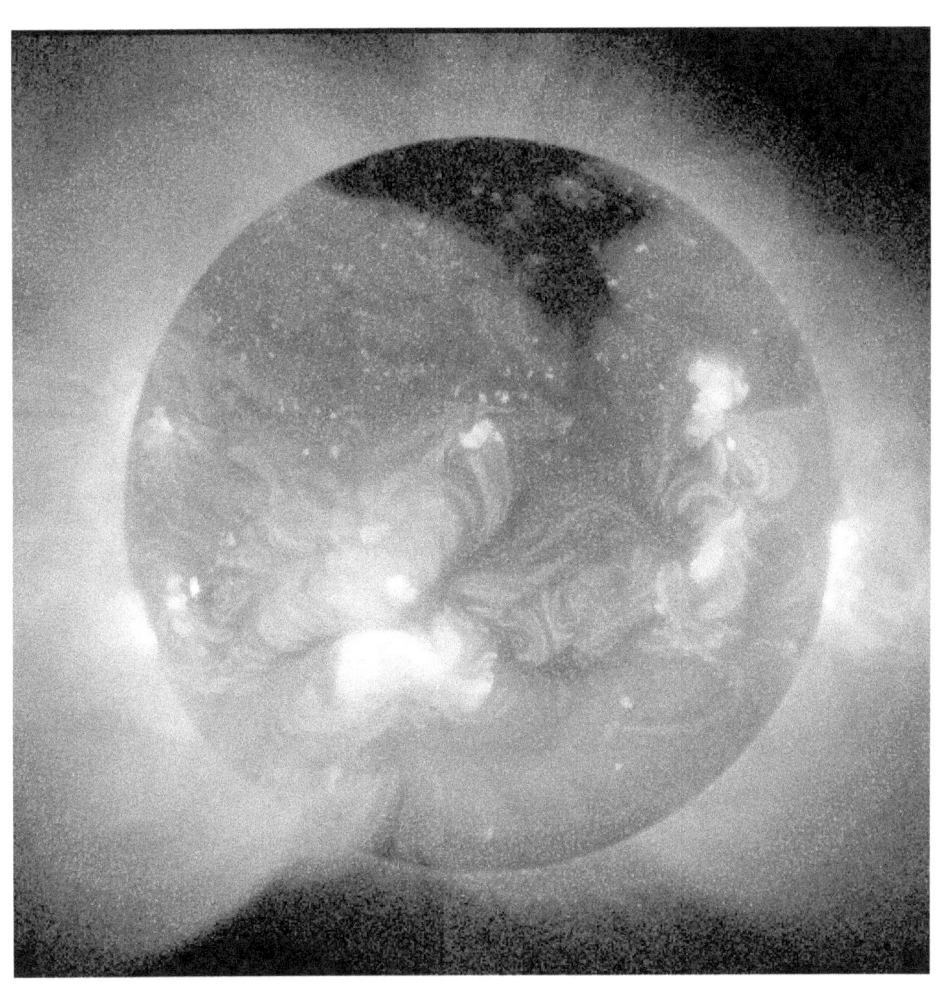

One of the meanings of the word "fuse" is to unite or join together.

Fusion (Fuse)

The Sun is hot but it is not on fire. The secret to why the Sun shines is fusion! Inside the Sun, with lots of heat and pressure, light hydrogen (H) atoms join together to make heavier elements of helium (He) and others.

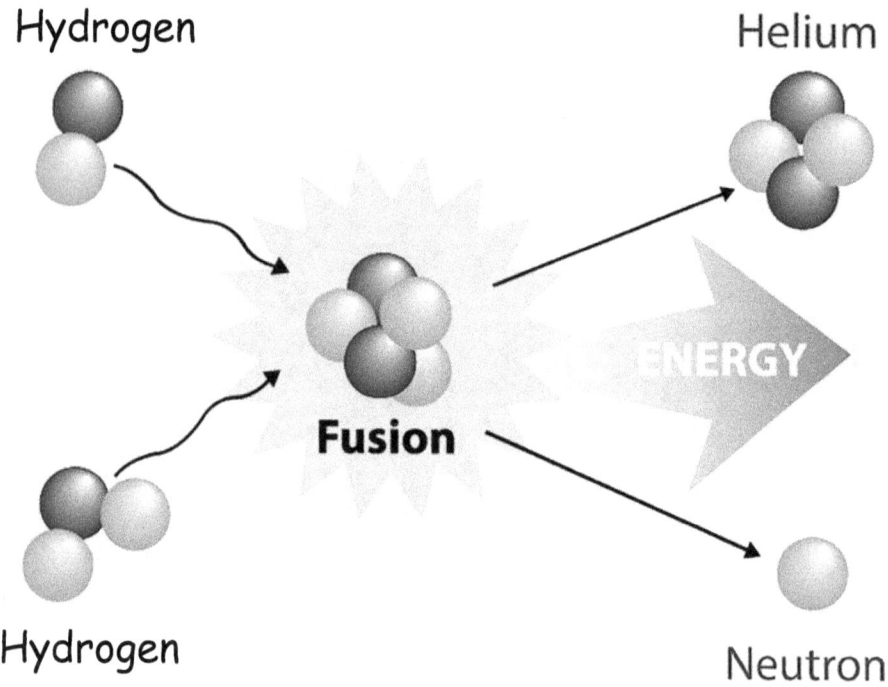

When atoms join together, energy is given off. This includes heat and light.

Seven Ideas

Fission (Fizz)

Interestingly, energy is also given off when heavy atoms break apart. This is called fission. Heavy uranium atoms break into lighter atoms and energy. This is the power source of atomic bombs and nuclear energy that makes electricity.

Uranium

^{236}U

Krypton

Barium

^{92}Kr

^{141}Ba

energy

atomic bomb

Nuclear power plants use these splitting atoms to heat water into steam. The steam then turns giant turbines to make electricity.

This is a reminder that science can be used for negative or positive results. This book is all about applying science to improve lives.

2) Electricity and Magnets are Related.

a) Induction

b) Electronics

c) Electric Light

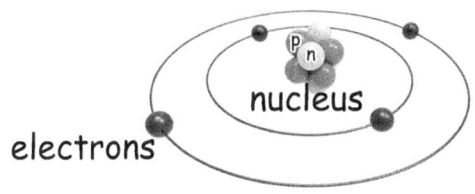

We start with our friend the atom. Each atom has a central nucleus surrounded by negatively charged electrons.

2a) Induction

Just how is electricity made?
Simply move a magnet near a wire
to cause the electrons to move.
Electricity is made of moving
electrons. That's it!

Falling water, coal fires, wind
and nuclear energy are all
used to turn wires near
magnets to make electricity.

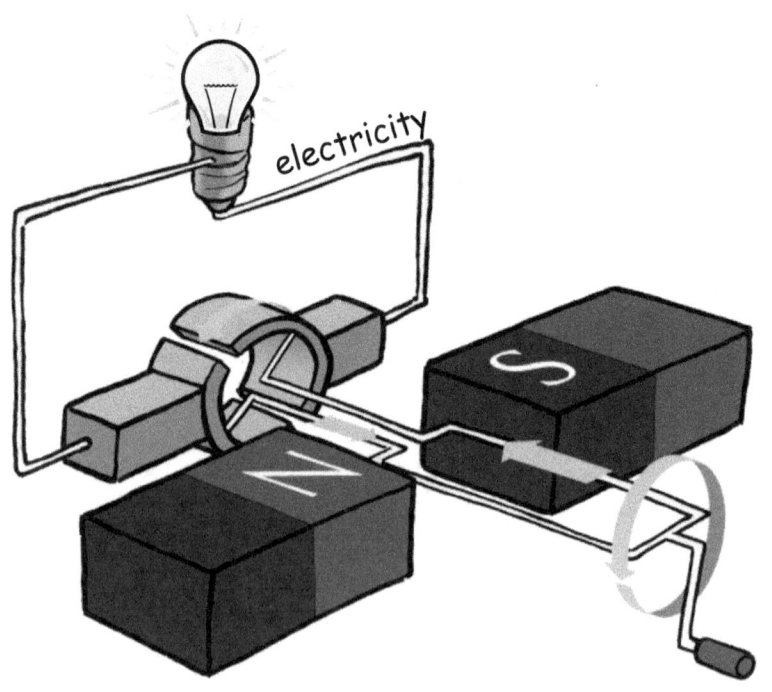

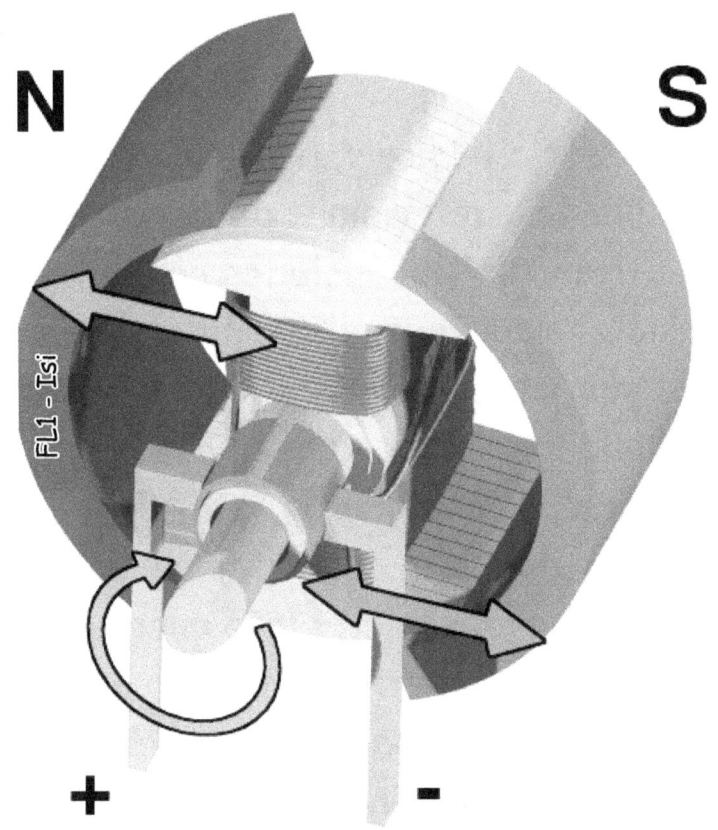

Electric motors work in reverse. Electricity flows through wire coils. This makes an inner magnet that pushes against the outer magnets to make the motor move. Magnets have two ends or poles. Just like how the Earth has north and south poles. With magnets, the same poles push apart and opposite poles attract each other. It is the flow of electrons that turns coils of wire into magnets. Electricity and magnetism are related.

2b) Electronics

As electricity flows, it also powers telephones, televisions and other electronics. This includes computers. Inside computers, electricity flows like mini bursts of lightning. They flow along the linked lines of microchips to compute data.
Data is stored in computer memories by turning electricity into micro magnets. When the computer needs the data again, magnetic patterns are turned back into electricity.

Electricity can also be very bright.

2c) Electric Light

Electricity flows through a thin wire
called a filament to make a light bulb glow.

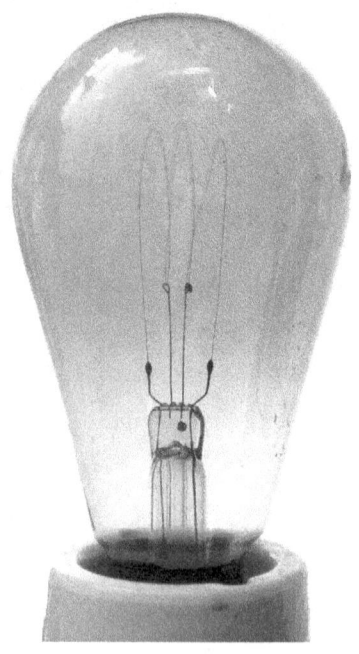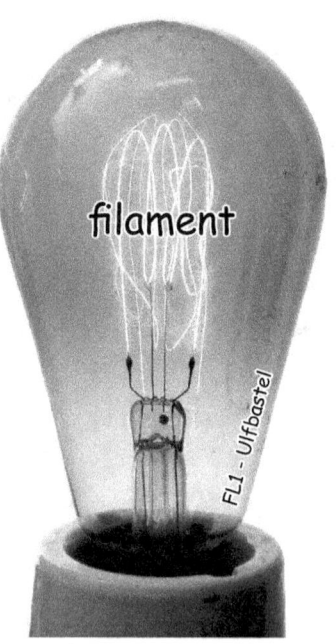

filament

FL1 - Ulfbaste

Electricity off.   Electricity on.

Seven Ideas

The electricity heats the thin
wire until it is so hot that it glows.

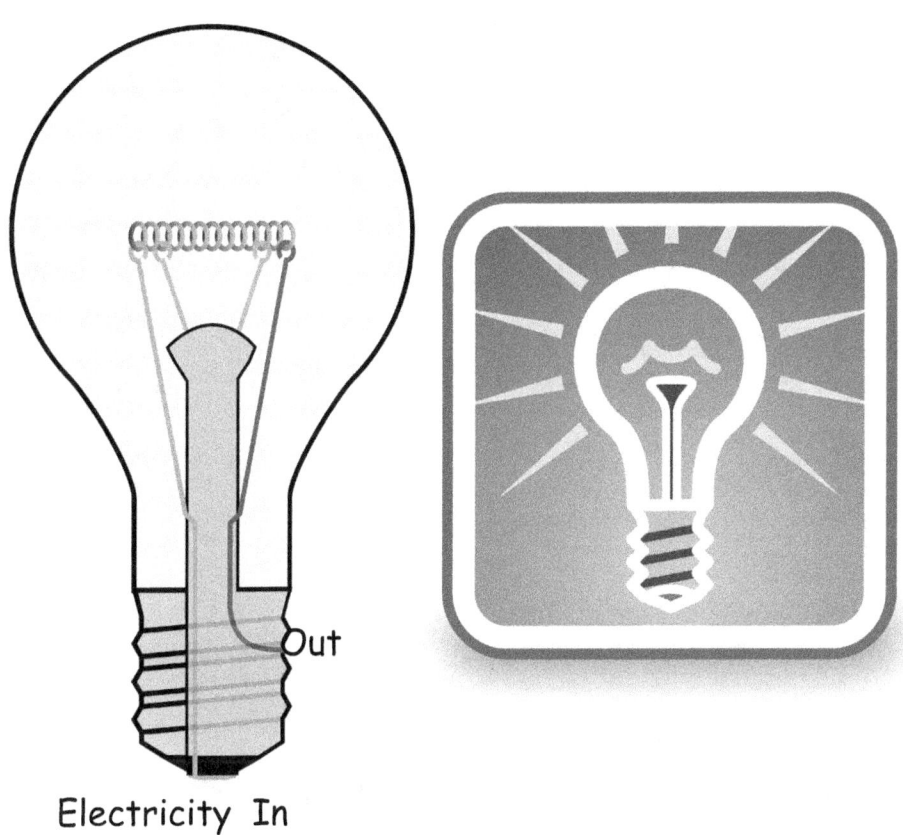

Out

Electricity In

This type of light bulb is called incandescent.

This is another type of light
bulb. Electricity heats these
mercury atoms until they make the
white coating of the light bulb glow
This gives off light.

These light bulbs use less electricity
but give off more light. These
fluorescent bulbs last longer
than the filament light bulbs.

Visible light

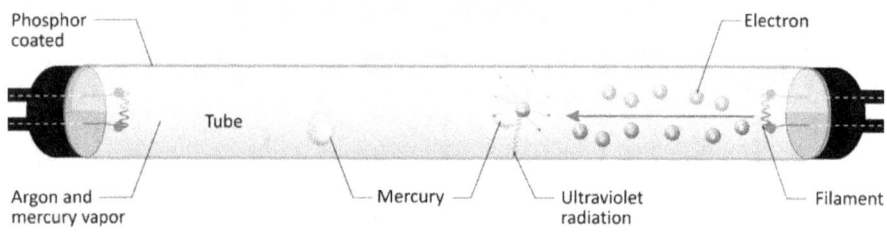

Phosphor coated

Electron

Tube

Argon and mercury vapor

Mercury

Ultraviolet radiation

Filament

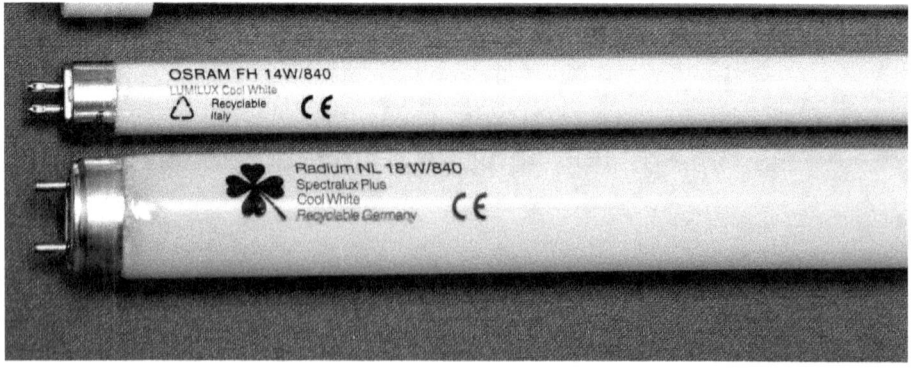

OSRAM FH 14W/840
LUMILUX Cool White
Recyclable
Italy
CE

Radium NL 18 W/840
Spectralux Plus
Cool White
Recyclable Germany
CE

Seven Ideas

Lights do more than just turn on and off.

3) Light can Bend, Bounce and Beam.

a) Refract

b) Reflect

c) Shine

Seven Ideas

It is hard to explain just what light is. It is easier to see what light can do.

FL1 - Cepheiden

Light bends or refracts through
a prism. The white light separates
into the different colors of a rainbow.

Seven Ideas

3a) Refract

Light bends (refracts) through lenses to focus images in eyes and cameras.

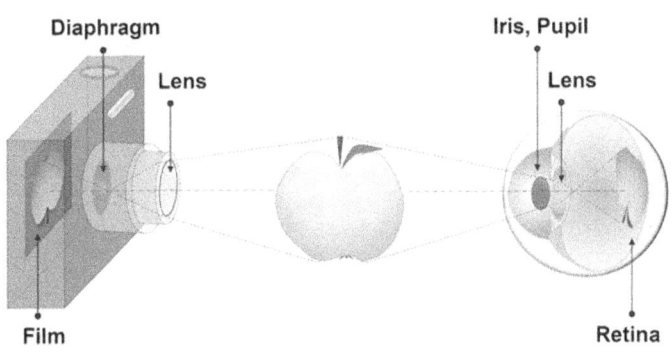

Diaphragm   Iris, Pupil

Lens   Lens

Film   Retina

Here is an example of how light bends.

The pencil is really straight but it looks bent because light refracts in the water.

Light also bounces.

Seven Ideas

3b) Reflect

You can see your reflection because light bounces off mirrors. Similarly, the Moon and planets don't have their own light sources. We see them because they reflect light from the Sun.

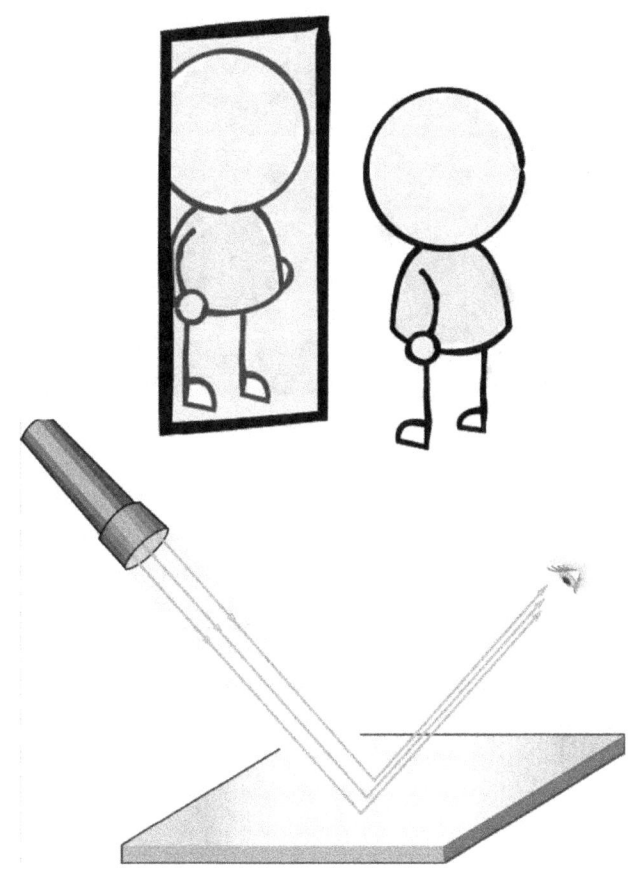

In this picture, sunlight bounces off the lake to reflect the mountain and trees.

The Sun is the main source of light.

3c) Shine

To recap, the Sun is not on fire.
Hydrogen (H) atoms join into
helium (He) atoms inside the Sun.
Sun fusion gives off energy and
light. We can see sunshine and
feel sunbeams on Earth.
Scientists think the Sun is over
four billion years old. It is
believed to be middle aged.
So, hopefully, it will shine for
at least four billion more years.
The Sun is very hot.

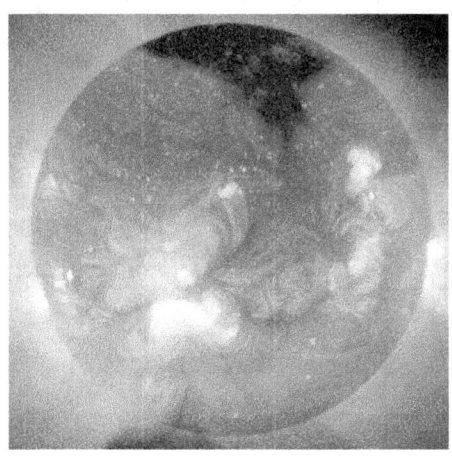

Light can be cool, too. Some creatures like fireflies and jellyfish make their own light. Inside their bodies, special cool chemicals mix together to make light.

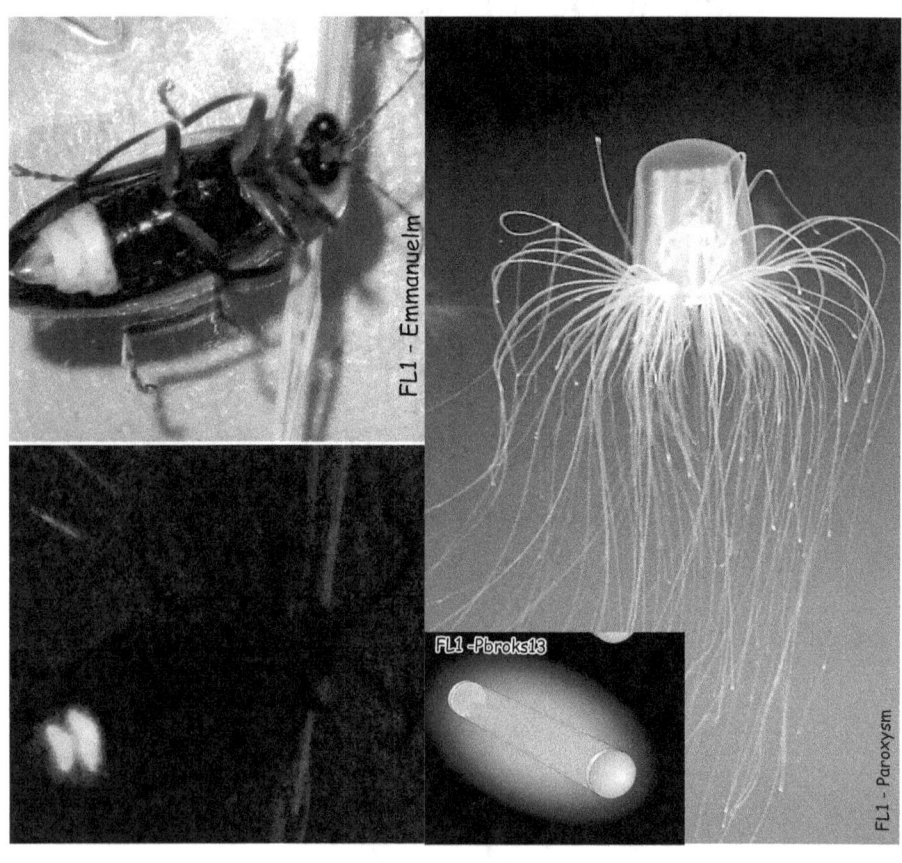

Light made by life is called "bioluminescence."

Glow sticks work the same way.

Light is also entertaining.

At the movies, light shines from
the projector's light bulb. Light
focuses through lenses, bounces
off the screen and into your eyes.

FL1 - Mattia Luigi Nappi

When you watch movies on electronic
screens, electricity turns into lights that
energize color scenes into stories to connect with.

Some movies are about dinosaurs.

Fair Use Jurassic World NBC Universal

Seven Ideas

4) Life Changes

a) Evolution

b) Fossils

c) Fossil Fuels

millions of years

Fair Use - David Gifford

4a) Evolution

Over time, life changes or evolves. Fossils show part of the story about ancient life. However, living animals are different from their distant ancestors.

Mammoth

Elephant

Seven Ideas

As fascinating as fossils are, they are only glimpses of how the world once was. Before fossils were discovered people thought that the world had always been the same. Instead, fossils show how life changes or evolves over time.

Ammonite   Nautilus

4b) Fossils

Before they became part of rocks,
creatures like dinosaurs were
actually alive. Fossils show interesting
life forms. There is awe and curiosity
about what the world was like back then.

Why is oil called a fossil fuel?

4c) Fossil Fuels

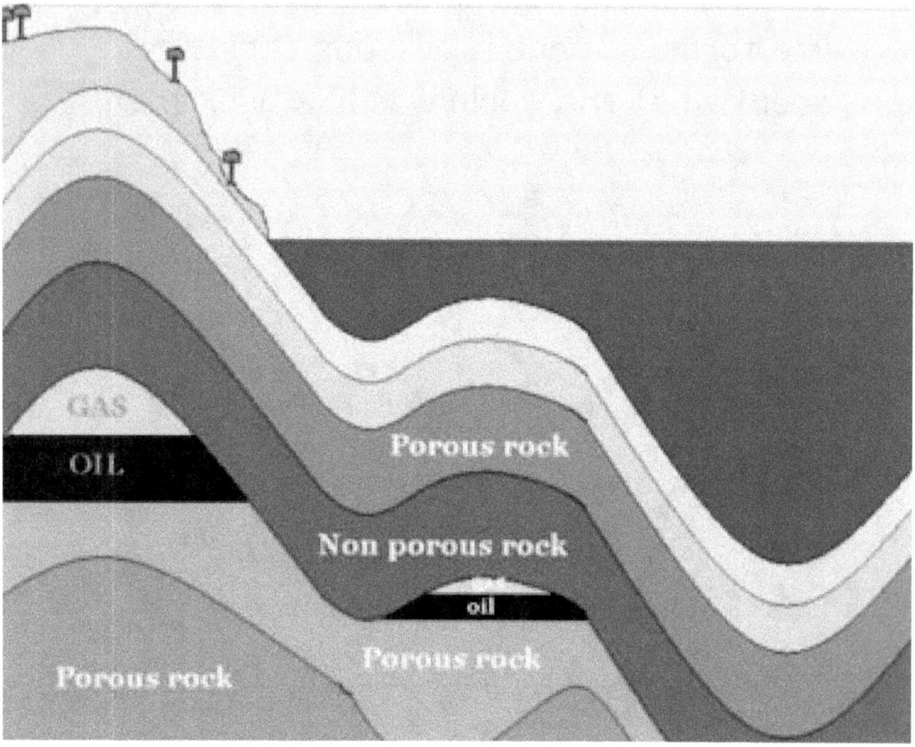

GAS

OIL

Porous rock

Non porous rock

oil

Porous rock

Porous rock

Everything that was and is
alive has carbon (C) atoms.
Let's go back in time
to millions of years ago.

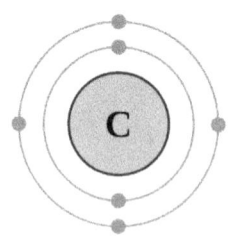

Ancient ocean life - like plankton and fish - died. It settled on the bottom of the shallow seas. The layers of carbon and hydrogen remains were covered with sand.

Over time, heat and pressure turned the carbon into oil and natural gas.

Also, ancient land plants died and piled on top of each other to become coal.

This is why oil, natural gas and coal are called fossil fuels.

Over millions of years the earth changed. The ancient sea sand turned into buried rocks with pockets of oil and gas.

Oil seeps to the surface from far below.

How did our ancestors learn about oil?
Oil floats on water. Even though it is
buried deep underground, small amounts
of oil seep through cracks in the rocks
over time. Oil makes its way to the
surface. Ancient people used oil to
waterproof bowls and boats. They
also learned that oil burns.

Seven Ideas

Today, people drill holes into the earth to get oil and natural gas.

Oil and gas are used as fuel to get energy.

5) Energy Changes Form

a) Fuel Burns

b) Fire Triangle

c) Engines Move

5a) Fuel Burns

Wood has carbon (C) atoms.
Wood burns.

When fuel burns, it changes
chemical energy into the
heat and light energy of fire.

Since the time of living in caves,
humans have used fire to cook and heat homes.

Seven Ideas

5b) Fire Triangle

Fire has three parts like a triangle. The carbon (C) atoms in the fuel quickly combine with oxygen (O) atoms from the air. When the atoms join they give off heat and light.

Fire is also called "combustion."

To recap, a match adds heat
to fuel that combines with
oxygen to give off heat and light.

5c) Engines Move

Fire gives machines power.

Controlled bursts of burning gas push
pistons down to make car engines move.

In a jet engine, fuel
and air come together.
They burn and then the
exhaust quickly pushes
out.

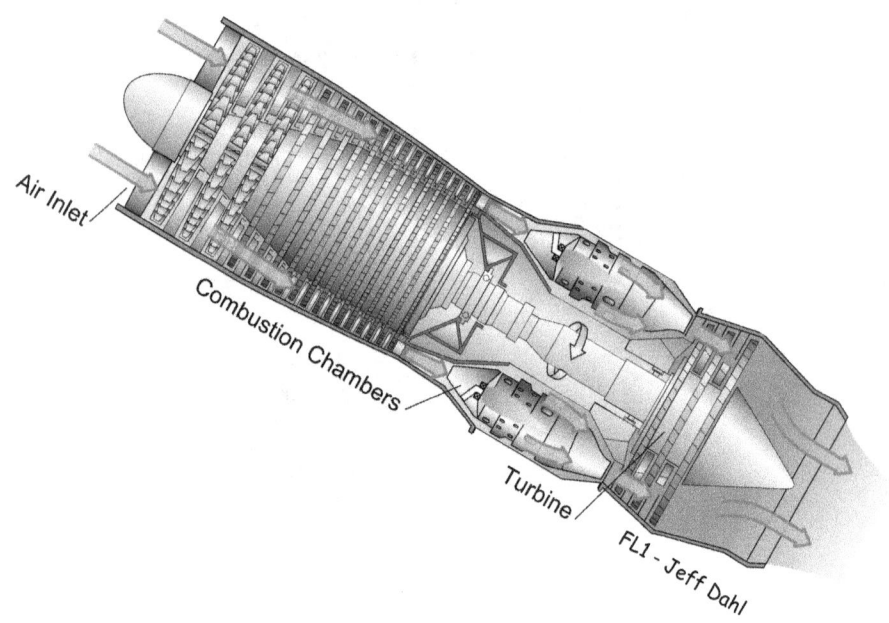

Air Inlet

Combustion Chambers

Turbine

FL1 - Jeff Dahl

It is the energy from the burnt
fuel that pushes the plane forward.

Radio waves help keep planes
from crashing into each other.

6) Radio Waves Are Useful.

a) Radar

b) Radio

c) Wireless

Radio Wave Dish

FL1 - Darren Kirby

Seven Ideas

6a) Radar

Radar sends out radio waves. Some of the waves reflect off of objects like airplanes. The reflected waves return to the radar station.

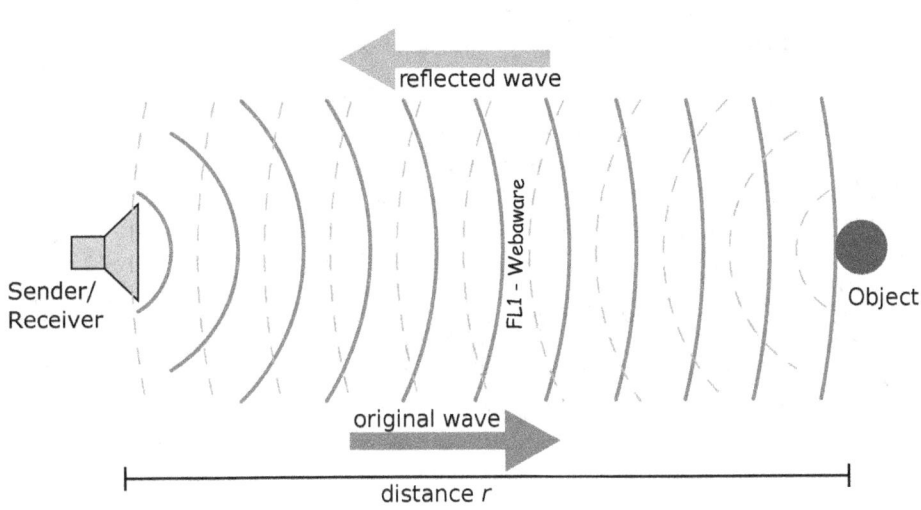

reflected wave

FL1 - Webaware

Sender/
Receiver

Object

original wave

distance *r*

Bounced radio waves turn into blips on the radar screen. They show the planes location and distance. Radar controls airplane traffic.

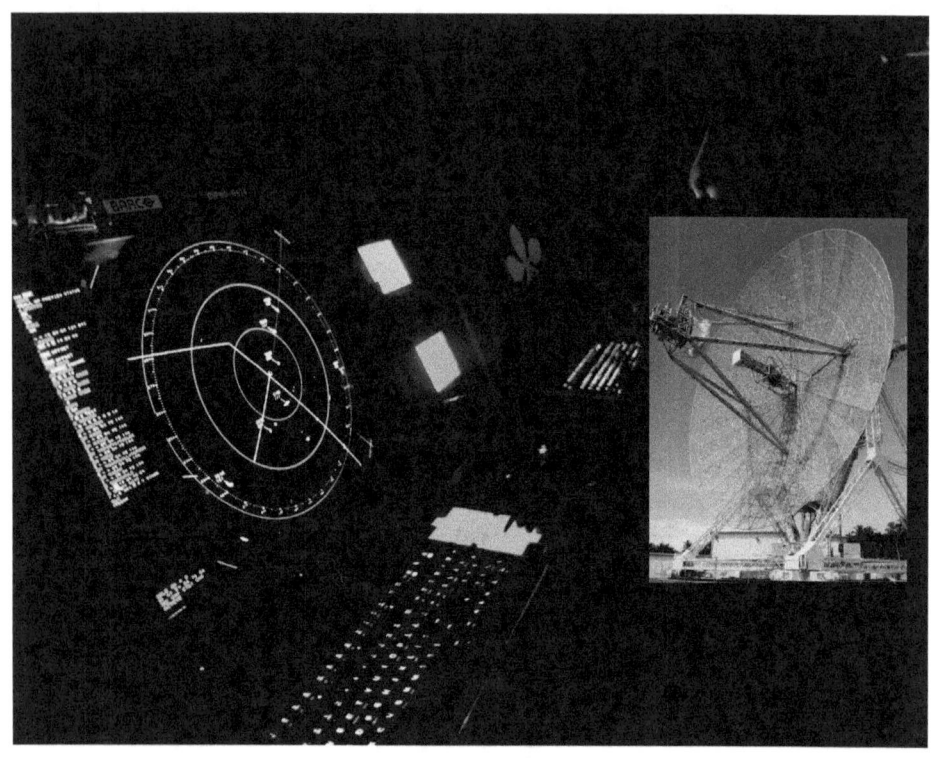

Planes and ground stations
use radios to communicate.

6b) Radio

Guess what radios use? Yup! They
use radio waves. Radio waves are
`cousins` to light. These waves are
called "electromagnetic" or EM.
They are made of tiny electric
and magnetic parts.

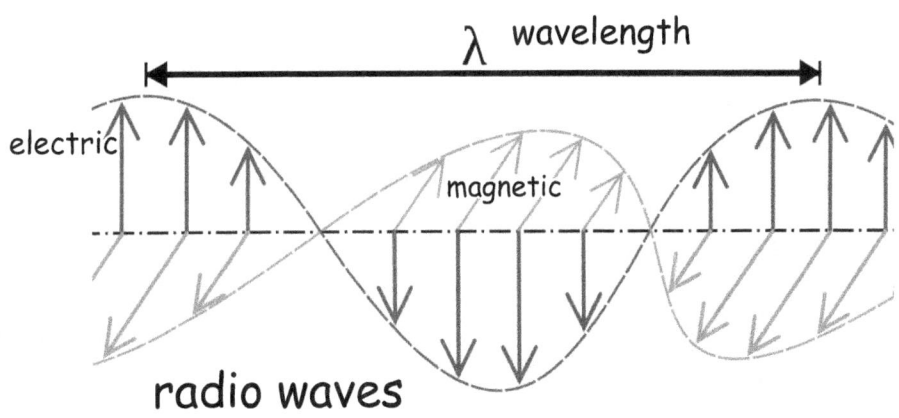

λ wavelength

electric

magnetic

radio waves

Different types of EM waves vary in length.
Radio waves are longer than light waves.
Light waves are the size of microscopic
bacteria.
Do you know that radio waves vary in length
from less than a meter to kilometers?

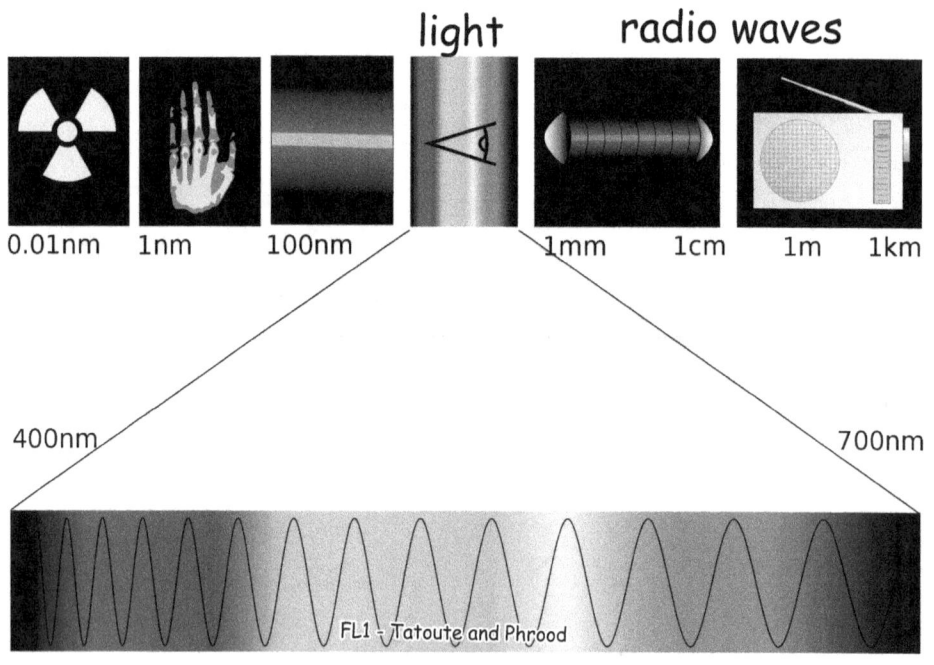

But how are radio waves made?

To recap, electricity makes magnets move in a motor! Quickly turning electricity on and off turns the magnets on and off too. Turning magnets on and off makes radio waves. Changing how fast the magnets go on and off changes the type of radio waves.

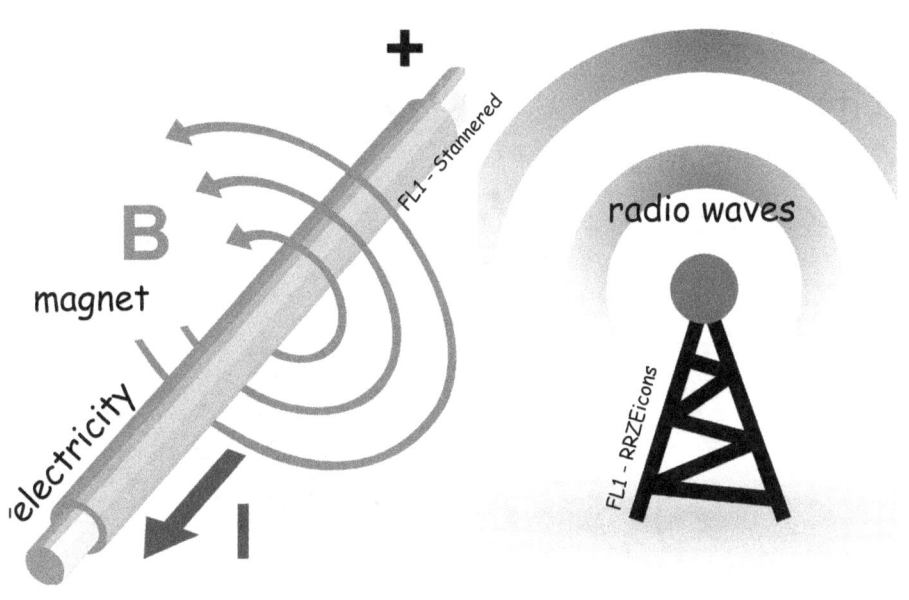

Radio stations have numbers based on the frequency or number of waves per second.

Radio waves flow from sending antennas. Radio waves flow through air and space. They go to the receiving antenna.
Car radios also work this way.

Radios don't need wires. It is why they are called "wireless."

Seven Ideas

6c) Wireless

WiFi is a type of wireless communication. It uses radio waves to connect computers together.

Smartphones also use wireless technology.

In a coverage area called a cell, there is a base station. Tower antennas send and receive your voice, texts and pictures using wireless radio waves. This is how our Apple and Android "cell" phones work.

7) Gravity is Attractive.

a) Apples Fall

b) Earth Orbits

c) Galaxy Glue

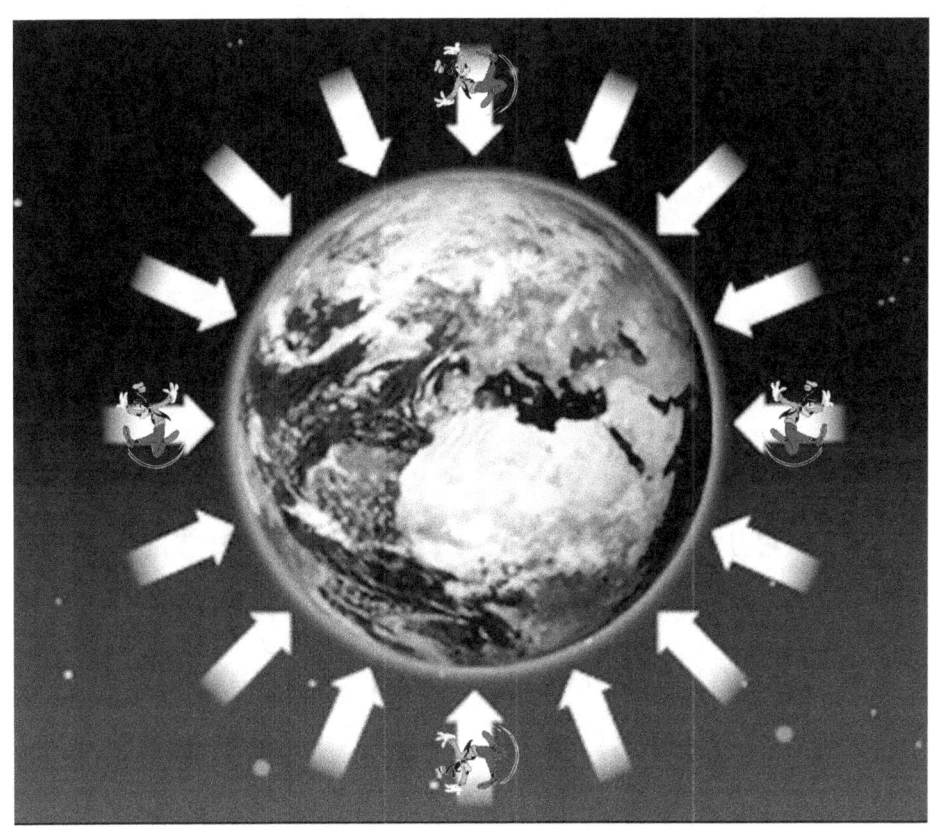

7a) Apples Fall

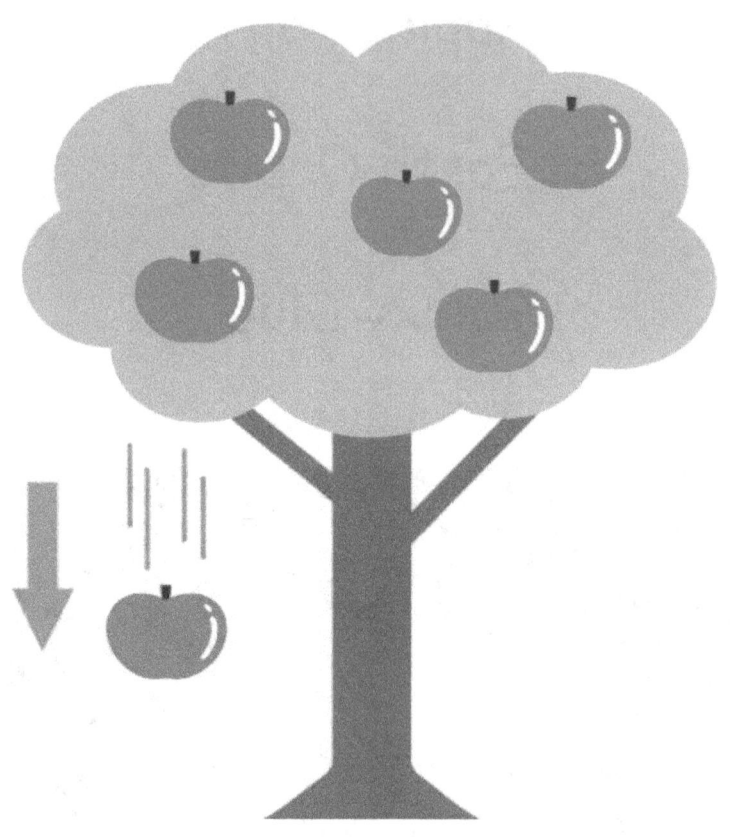

Real apples fall to the Earth because gravity pulls them down. Gravity is also what keeps our feet on the ground.

Gravity is the pulling power
where heavier objects pull
on or attract lighter ones.

7b) Earth Orbits

Gravity is why the Moon
circles the Earth. It is
also why the Earth goes
around or orbits the Sun.

All the planets in the solar
system orbit the Sun because
of gravity.

7c) Galaxy Glue

Our solar system is part of a galaxy that turns together in space because of gravity. Gravity is the "glue" that keeps it all together.

Milky Way Galaxy

Conclusion
— Revolve and Evolve

Our Milky Way Galaxy spiral turns or revolves together in space because of gravity. The galaxy echoes with distant radio waves that hint at the Big Bang origins of everything in the universe. Energy constantly changes form. Deep within stars, small atoms continue to join into bigger ones. They give off heat and light.

Stars shine like countless Christmas tree lights on the endless canvas of space all around us. From our Sun, light powers life on Earth, even as that life evolves.

Lightning strikes from sky to earth as electricity finds its way back to ground. The ground and all life on it are made of atoms. Over time, the ground is heated and pushed into rocks and mountains. Caves form. Caves are also symbols of where humans, our ancestors, once lived. Dark caves show us what life would still be like without science.

The essence of science is using knowledge like these seven ideas to better all of our lives! As science enables worldwide communications, may we embrace that we are all on this revolving and evolving adventure together!

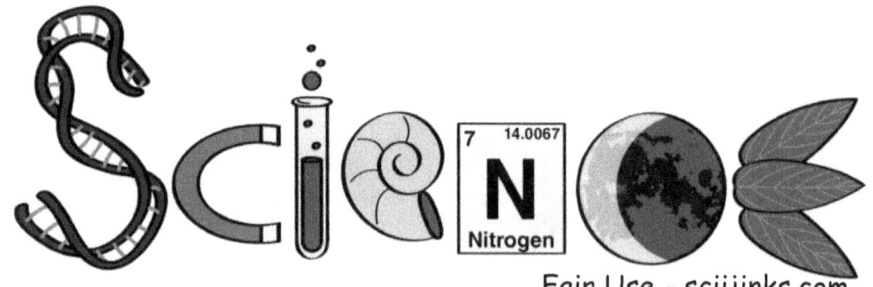

Fair Use - sciijinks.com

Seven Ideas

Credits

Unless otherwise noted,
pictures are in the public domain.
The following were utilized as resources
for research for this book:
. Note Flagnote or FL1 means Wikipedia at
www.wikipedia.org. Specific author is called out.
. Museum of Science & Industry in Manchester, UK
. London Science Museum
. NASA
. Kern County Museum California
. La Brea Tar Pits
Front Cover - Wikipedia Sergiu Baciou
Pg i - Wikipedia Lithium Atom by halfdan &
Milky Way Galaxy by NASA
Back Cover -Lightbulb by Stefan Krause, Germany
See http://commons.wikimedia.org/wiki/File:Glüh-
wendel_brennt_durch.jpg
Wikipedia Jellyfish by Paroxysm

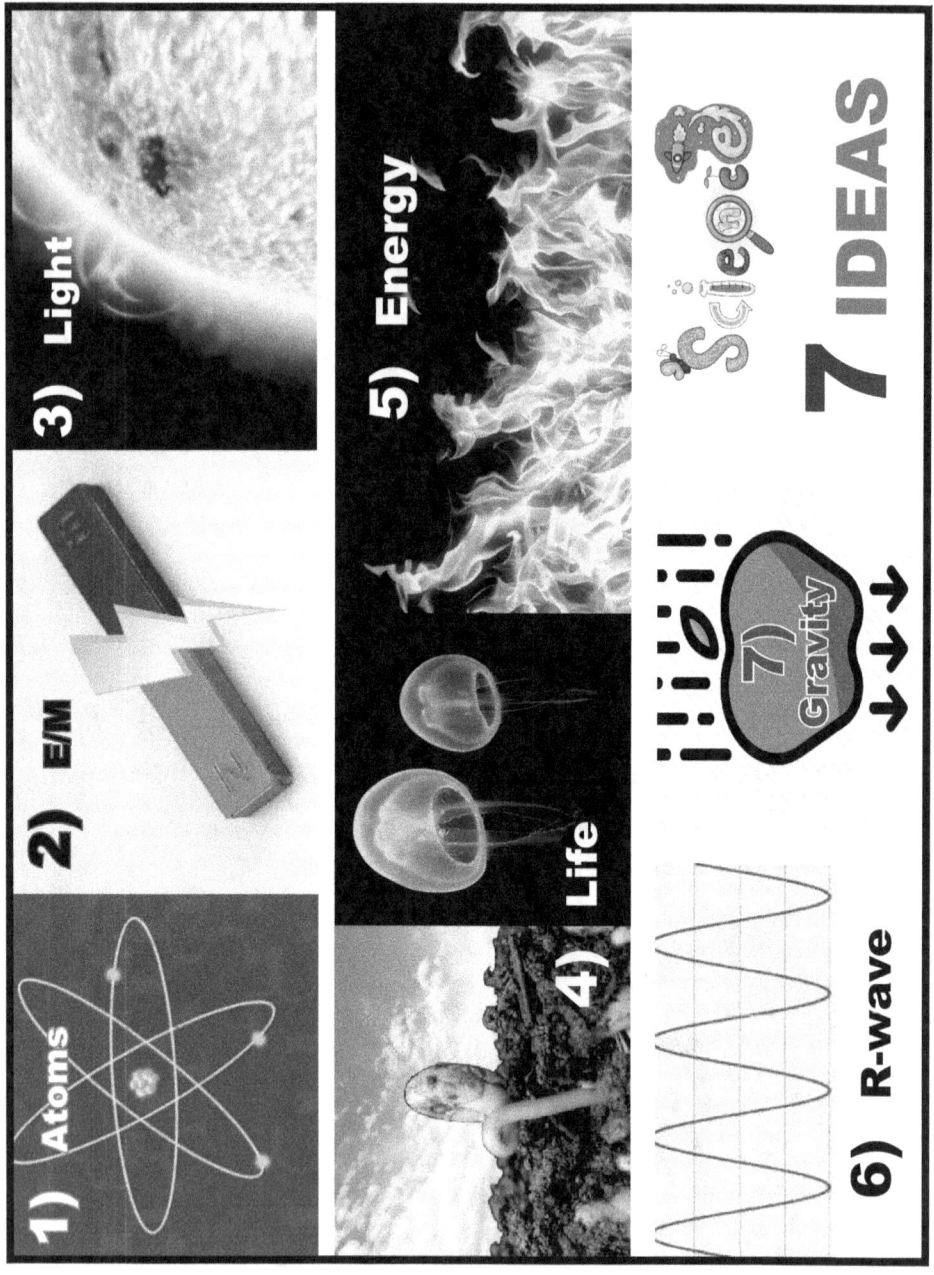

67

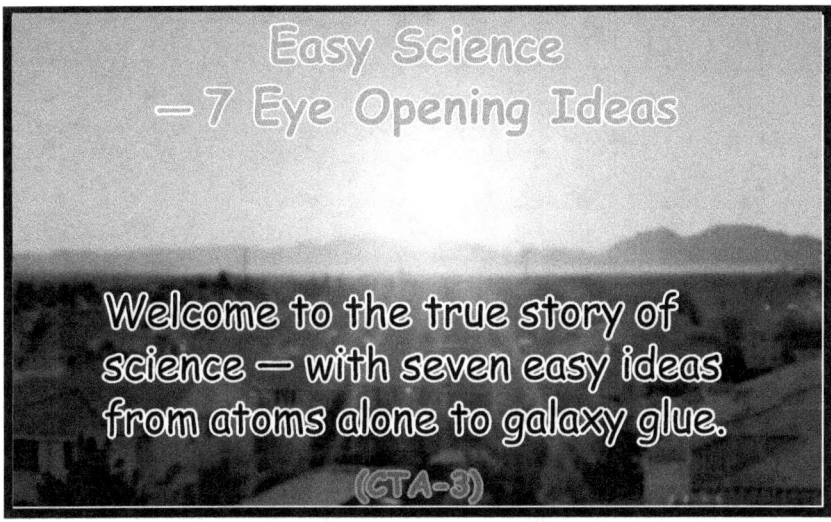

Easy Science
— 7 Eye Opening Ideas

Welcome to the true story of
science — with seven easy ideas
from atoms alone to galaxy glue.

(CTA-3)

For a copy of this video contact:

Main Points
Science is important to our lives!
1) Air flows in circles for seasons.
2) Water flows from high mountains to low seas and back again.
3) Food energy flows from the eaten to the eaters.
4) Energy flows from fuel to power planes, cars and rockets.
5) Heat flows from hot to cold.
6) Data flows to power our digital devices and cyber Silicon Age.
7) With DNA, lifeless atoms assemble into living cells.

For a copy of this video contact:

Seven Ideas

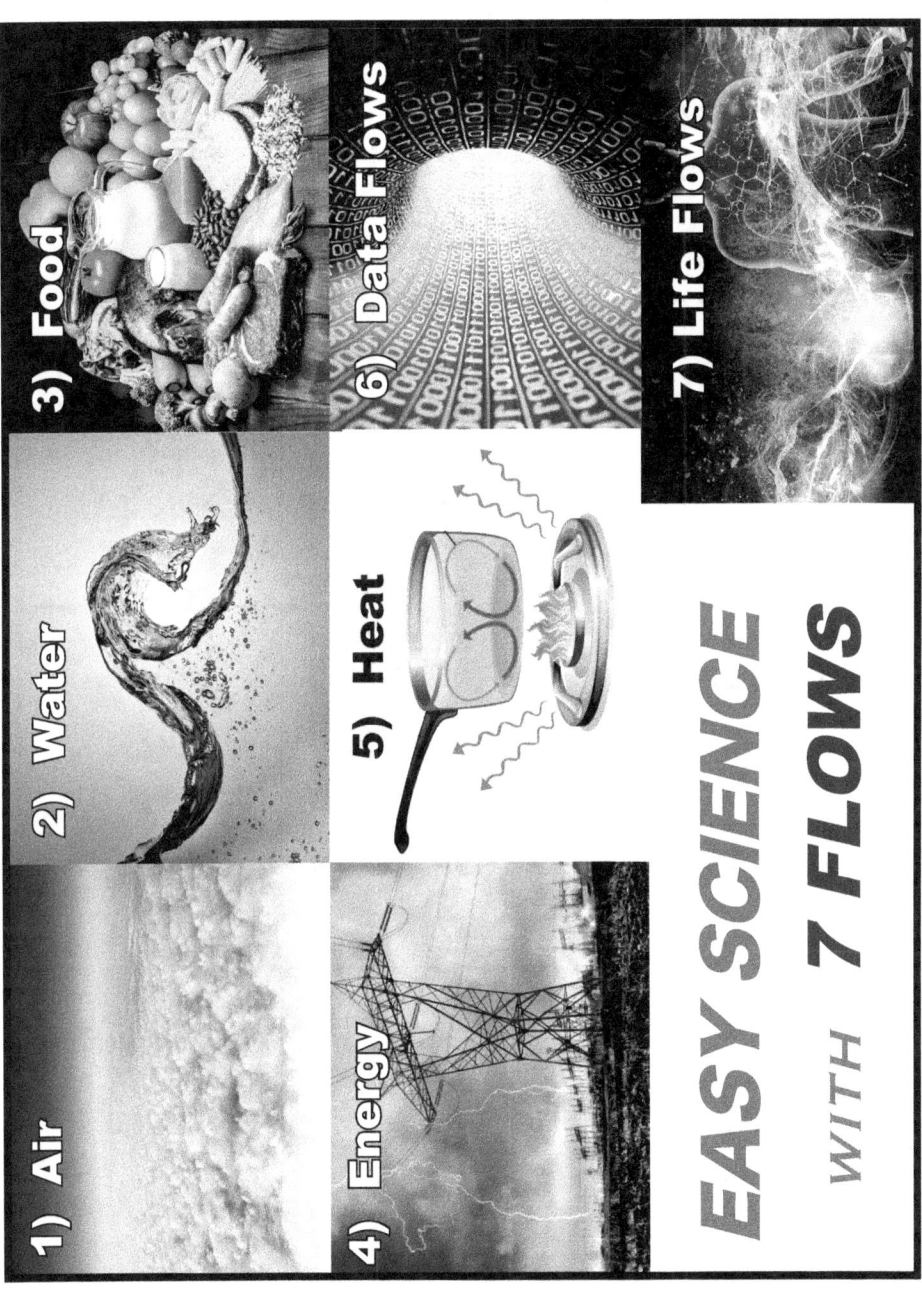

Easy Ideas	Airplanes	Cars
From Concepts to Critical Thinking	From Four Forces to Flights	From Actions to Autos
1	2	3
Computers	Smartphones	Food
From Digital to Data	From Calls to Global Connects	From Eats to Energies
4	5	6
Nature	Space	Light
From Atoms to All Life	From Elements to Us	From Suns to Sapiens
7	8	9
AI	STEM-Zen Program	Everyday Objects
From Machine Muscles to Minds	From Empty to Science EnLights	From Ideas to Daily Items
10		Bonus

Seven Ideas

Seven Eye — Opening Ideas

This is the simple story of science from atoms and oil to waves and wireless. This book explores seven easy-to-understand ideas. For it is science that brings light to night and improves our lives.

BONUS
Download FREE VIDEO (CTA-4)

Science is important to our lives!
1) Air flows in circles for seasons.
2) Water flows from high mountains to low seas and back again.
3) Food energy flows from the eaten to the eaters.
4) Energy flows from fuel to power planes, cars and rockets.
5) Heat flows from hot to cold.
6) Data flows to power our digital devices and cyber silicon age.
7) With DNA, lifeless atoms assemble into living cells.

73

7 FLOWS VIDEO Script

i-1) Welcome to Easy Science with 7 Flows

i-2) Every minute, life needs air. There are more atoms in a breath of air then the number of breathes of air on earth.

i-3) Water covers about 2/3rd of the Earth.
About 2/3rds of our bodies are water too.

i-4) Food is the fuel of animals and us!

i-5) Applied energy is power from the Sun and machines to ourselves.

i-6) Uneven heat is the source of our seasons

i-7) Data is the electronic heart beats to our digital devices.

i-8) Let's see, 7 flows of Lively Science!

(ONE PAGER without text, Air, Water, Food, Energy, Heat, Data, Life Flows)

1-1)   ONE, Air
Long ago, earth had no oxygen. (Hadean Eon)

1-2)   (Blue-Green Bacteria/Algae) Later, Life evolves that exhales oxygen.

1-3)   Over time, oxygen builds up around the world (text, atmosphere). (Picture N2 = 78%, O2 = 20%)

1-4)   Special microbes turn Nitrogen in the air into soil fertilizer that are plant vitamins (text, Nitrogen Fixation)

1-5)   (Forest) Fire burns oxygen. Animals and us breathe in oxygen. Our cells use oxygen to "burn" food bits into energy.

1-6)   Earth's atmosphere (Text, 100km thick) pushes on us with air pressure
(text, 14.7psi)

1-7)   Sun shines on the earth unevenly (show earth tilt and spin).
Air with different amounts of heat causes our seasons and weather. (Show storms and sunny days)

1-8)   Air moves clouds that are full of water too.

2-1) TWO, Water
The source of Earth's water may have been ancient comet crashes.

2-2) Earth's first life evolves in the oceans.

2-3) The sun powers the water cycle. Sea water evaporates. Clouds move over land. Rain falls and flows down into rivers and back to the sea. This circle (water cycle) waters our planet, plants and people along the way.

2-4) Different sun heat causes different states of water.
(show ice, liquid water and steam)
2-5) Things dissolve in water. This is critical to life
(show plants, animals and people drinking).
Cells need water to function.
We also use water with soap to keep us clean.
2-6) Water can wear away mountains (text, erosion).
Overtime, flowing water made the Grand Canyon.
2-7) With gravity, water flows from high to low places.
We see this in our homes (show tap, use, drain)
2-8) Water helps us prepare our food too.

3-1) THREE, Food
Food is chemistry in action for life.
3-2) Solar powered plants, turn sunshine into sugars (text carbs).
Plants store energy as carbs like fruits and veggies.
3-3) Animals eat the plants. Animals turn carbs into energy.
Animals store energy as fat.
3-4) Omni eaters like people eat both plants (text, carbs) and animals
(text, fats and proteins). We get carbs and vitamins from plants. We
get fats and protein from meat.
3-5) Much of our bodies are made of protein. Here are some examples. (muscles, organs, tissues?)
Cells use proteins (show amino acids) like work orders to do things
like: digest food; (show one -amylase,lipase, pepsin), help us grow
(hormones) and fight germs (show antibodies).
3-6) Humans have about 5 Liters of blood cells.
Every day, blood flows in a circle around our body more than a 1,000
times.
3-7) Nature is full of makers (plants) and takers (animals).
All life is connected in a food web from microbes to me.
3-8) We get our energy by eating food.

4-1) FOUR, Energy
Plants change sunshine into chemical energy.
4-2) Fire changes chemicals in fuel into flame energy.
(Text, combustion)
4-3) Fire and Fuel power: planes; trains and cars.
Rockets too!
4-4) With Gravity, high things have potential energy.
This is why, mountain water wants to flow to the sea.

4-5) Moving things have kinetic energy.
We design cars carefully because they can crash.
4-6) Some big storms (text, weather) have attitude and energy like hurricanes and typhoons.
4-7) Electricity and magnets make electrical motors move.
(motors in fridges, washing machines and air con)
Also, electricity powers our lights and electronic (digital) devices.
4-8) Energy is the oomph that makes things move.
From muscles, motors and more (examples)
Energy is often hot!

5-1) FIVE, Heat
Heat is the flow of energy from Hot to Cold.
5-2) Temperatures measure how much heat energy objects have.
5-3) Here are three ways that heat flows: Rays, Touch and Circles.
5-4) Sun Rays (text, EM Radiation) shine through space
to warm our Earth.
5-5) Heat flows when things touch. (Text, Conduction)
Like (gas) flames to metal pots and pans to cook our food.
5-6) Heat flows with circles in air and water. (Text, Convection)
Like when we boil water. (Show convection currents)
Furnaces heat and blow air around our warm homes.
Heat flows in other circles in our cool fridges and air cons.
(show compressor cycle).
5-7) (Show Earth Tilt) Sunshine heats the earth unevenly. This causes hot and cold seasons. Heat flows in Air circles (text, Coriolis Effect) to make our weather. Currents of water flow from hotter to cooler places too. (show English current)
5-8) We collect data to understand Nature's Storms. (Example, typhoons, tornadoes, monsoon, Snow Blizzard etc)

6-1) SIX, Data Flows
We see how data flows when our senses, sense our world.
We send signals to our brains. We think and then tell our muscles to move.
6-2) We input (type on keyboard, scan, USBs)
data into our computers.
6-3) Like brain cells, our (computer chips) digital devices
are powered by (text transistor) gates and flows of e-bit pulses.
Data flows like clockwork driven by electrical heart beats.
6-4) Inside our electronics (examples), we assemble data into info outputs. We send texts, write (documents and) books, hear (music) sounds and see videos.

7 FLOWS VIDEO Script

2All this is cyber sync's into real-time parades of multimedia digital data flows.

6-5) This data changes from electricity bits to magnetic storage to light from screens.

Data flows, fluently among these different forms of energy. Constantly, the dynamic data changes form and then back again.

6-6) For examples, think of the dance of our digital data, when we use our smartphones. Data goes from e-bits to RF Radio Waves to colorful displays (screens) full of fun digital meanings.

6-7) The front face of our reality is defined by interfaces with real and virtual data flows. Where cascades of cyber codes augment and enhance our on-line lives.

7-1) SEVEN, Life Flows

It is amazing how my DNA changes lifeless atoms into full-of-life me.

7-2) I interface with my environment for the things I need (show air, water and food).

7-3) Life chemistry flows when I turn air, water and food into my energy (text, respiration).

7-4) Think of the complexity, as my tens of trillions of separate cells, all work together to keep my life going.

Blood flows (with Air, Water and Food Bits) to keep me alive.

I breathe, drink and eat. (Show breathing, circulation and digestion systems.)

7-5) Same is true of every life form. (text, biodiversity)

7-6) We do Not own our atoms. One day all life dies including us. Our well-used atoms return to nature - ready for reuse. (show decomposers)

7-7) While we are alive, May we marvel at our self-thinking minds.

As we think about the miracle of how Science flows with Life.

C-1) Now to Close! Air flows in circles that make our seasons.

C-2) Water flows from high mountains to low seas and back again (River Water flows to Seas. Evaporation clouds)

C-3) Food flows from the eaten to the eaters that connect the world wide Food web.

C-4) Energy flows from fuel to our planes, cars and rockets.

C-5) Heat flows from hot to cold.

C-6) Data flows to power our digital devices and cyber silicon age.

C-7) With DNA, lifeless atoms, assemble into live cells.

C-8) With 7 Flows, Science is Awesomely important to our lives!

(ONE PAGER with text, Air, Water, Food, Energy, Heat, Data, Life Flows)

Power
— Windows to Wheels

Science links POWER
from making glass
windows, then boiling
beer to connected
steam engines.
This impacts our world
as people change from
being farmers to
working in steam-
powered factories.

Power
— Windows to Wheels

Table of Contents

Without power,
muscles lift within limits.
With science we understand power
and how to move machines and ourselves.

POWER, One Pager

Let's see how science connects from glass windows to iron wheels.

In this chapter, watch for the following science principles.

- Heat
- Fermentation
- Decomposers
- Combustion
- States of Matter
- Smelting
- Pressure
- Power
- Acceleration
- Electricity Generation

Welcome to the true
tale of power that
links to our technology today!

With science, we progress from
glass windows to iron train wheels.

This is the true story
where the sands of time or
at least sand becomes glass.

Coal fuels fires. Also, expensive brass objects are replaced by practical iron ones.

Where heated water becomes steam...

Power

That powers an Industrial Revolution.

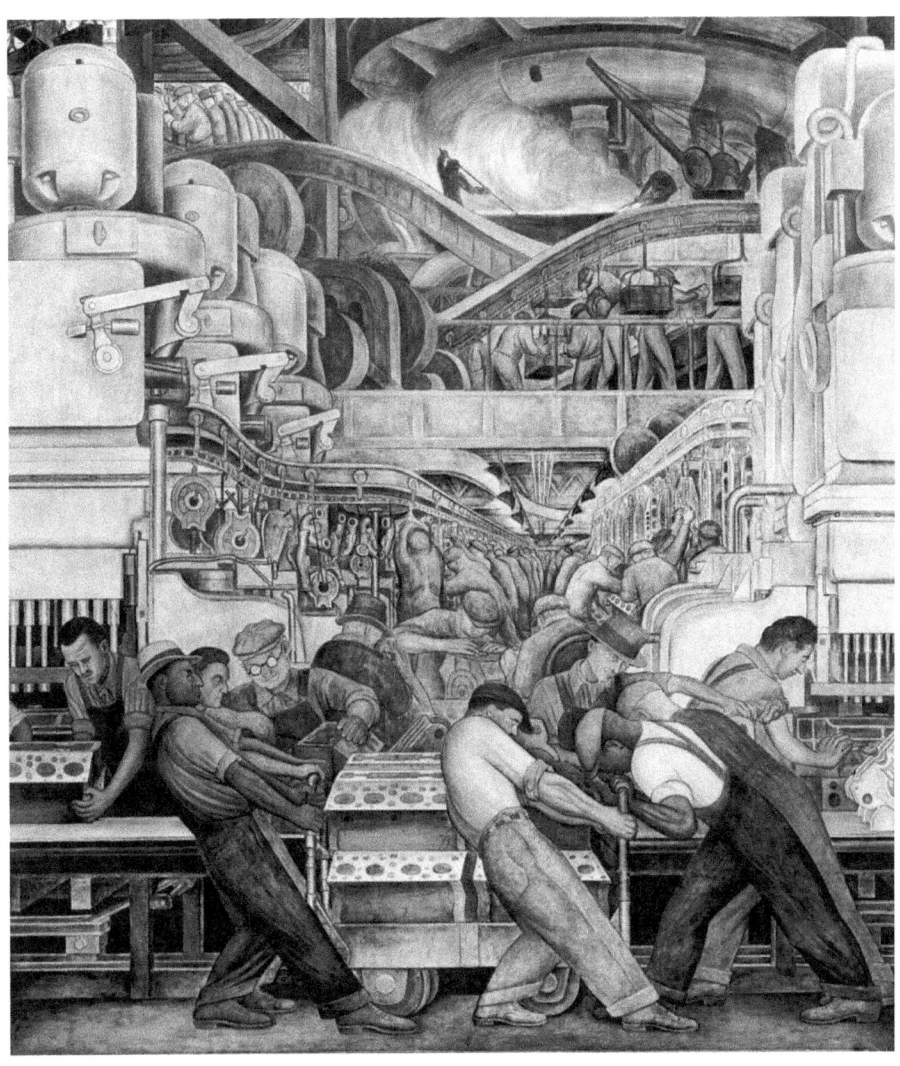

Power

There are links from the past to our present tech objects.

Let's start our quest
— from windows to steam
powered wheels — with a
substance we can see through.

ONE, **Blow Windows**

For thousands of years, glass
is made by melting these
together with lots of heat.

Glass is made of sand, soda,
lime and smashed glass
pieces called "cullet."

Long ago, the Romans use pipes
to blow glass to make lots
of cups, bottles and bowls.

Later, people learn to blow a big glass bubble and spin it. The bubble collapses into a disk shape. The flat, round crown glass is cut into small pieces. Lead joins the glass parts into window panes.

Crown Glass

crown glass manufacture, C18th
Diderot, *Encyclopédie*, sv 'Verrerie', pl 13

Power

Next, people make hot, huge cylinders of blown glass to make larger windows.

Today, window glass is made
by floating on hot tin but we will
save that story for a different day.

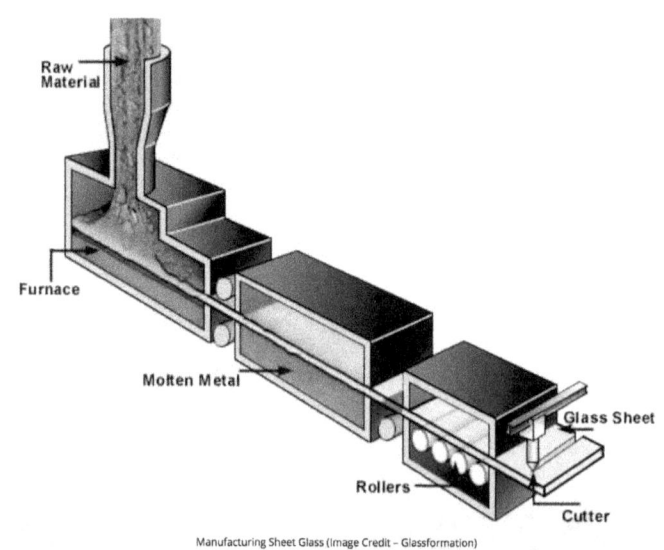

Manufacturing Sheet Glass (Image Credit – Glassformation)

Hundreds of years ago,
at a glass factory workers drink
beer to keep cool. Also, it helps
the workers not get sick from the
smoky fires that melt the glass.

What is the beer made of?

Ingredients

barley   hops   water   yeast

How is beer made and
how does this connect to our tech today?

Power

TWO, **Brewer Boilers**

Beer makers mix together grain, hops, water and yeast.

Water

Grain

Hops

Yeast

Next, the mash is boiled in giant copper vats called kettles.

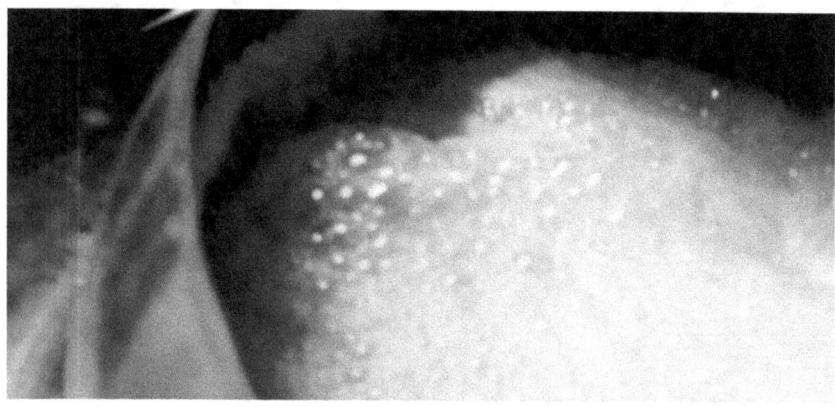

Fermentation

Yeast, a fungus, changes sugars in the mashed grain into alcohol.

Later, finished beer is put into glass bottles.

At first, the beer and glass bottle makers use wood as fuel.

In Europe, so many people burn wood,
the forests almost run out of trees.

Many businesses, like beer and bottle makers, start using coal as fuel.

Where does the coal come from?

THREE, **Coal Burns**

We go back hundreds of millions of years ago, before the dinosaurs.

This is a time of amphibians and giant insects called the Carboniferous Period.

Massive plants grow
in the hot rain forests
that cover the land.

When the huge plants die, they pile up on top of each other in layers called peat.

Peat

For "Pete's sake" or is that
"Peat's sake," over time, the
layers of plant peat get covered up
with sand. With heat, pressure
and time, the peat turns into coal.

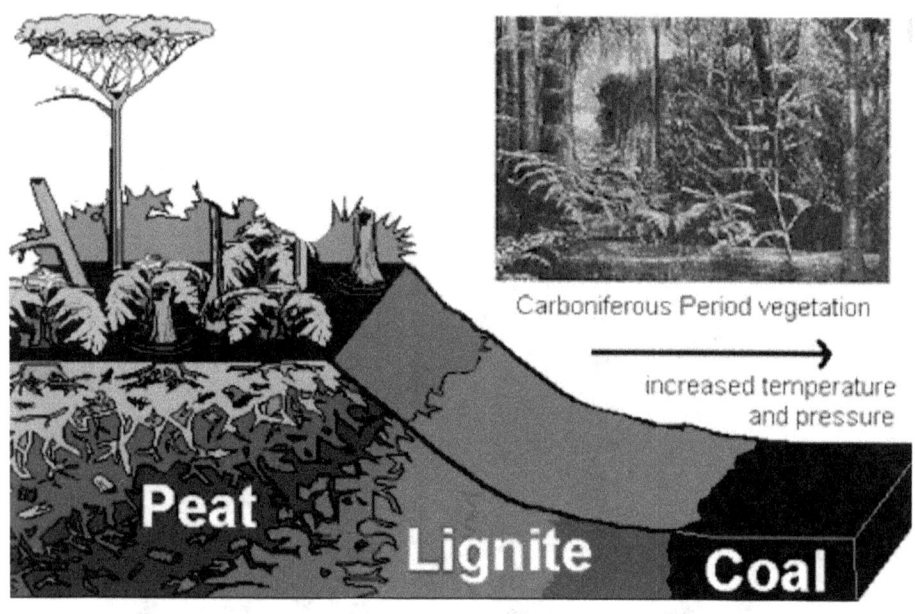

Carboniferous Period vegetation

increased temperature
and pressure

Peat

Lignite

Coal

At this time, the hotter, wetter climate produce
more plant matter than decomposers can deal with.
Maybe all the decomposers have not evolved yet.

| Worm | Mushroom | Insects | Bacteria |

Over hundreds of millions of years, the Earth churns. Some of the once deep buried coal finds its way to the surface. People find the coal and discover that coal burns. People use coal as fuel.

Coal

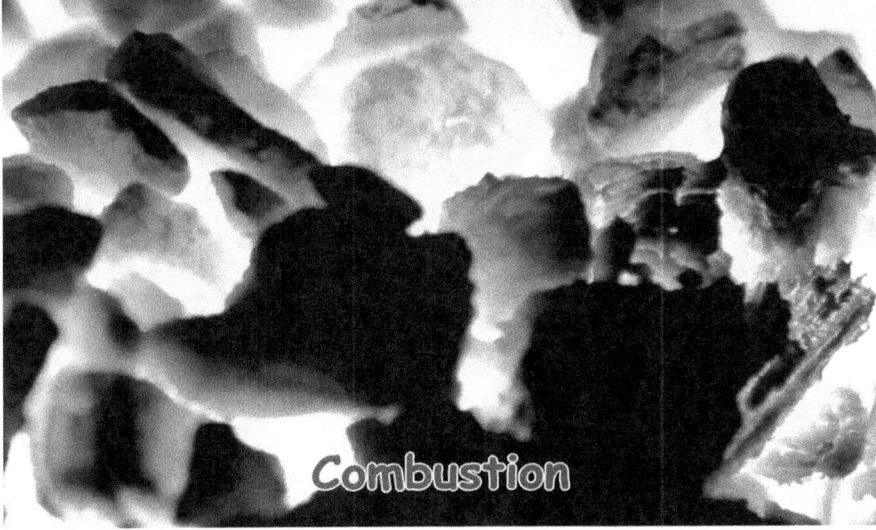

Combustion

Miners dig deeper
and deeper into the earth
to get coal and metal ores.
There is a big problem!

The mines flood with water.

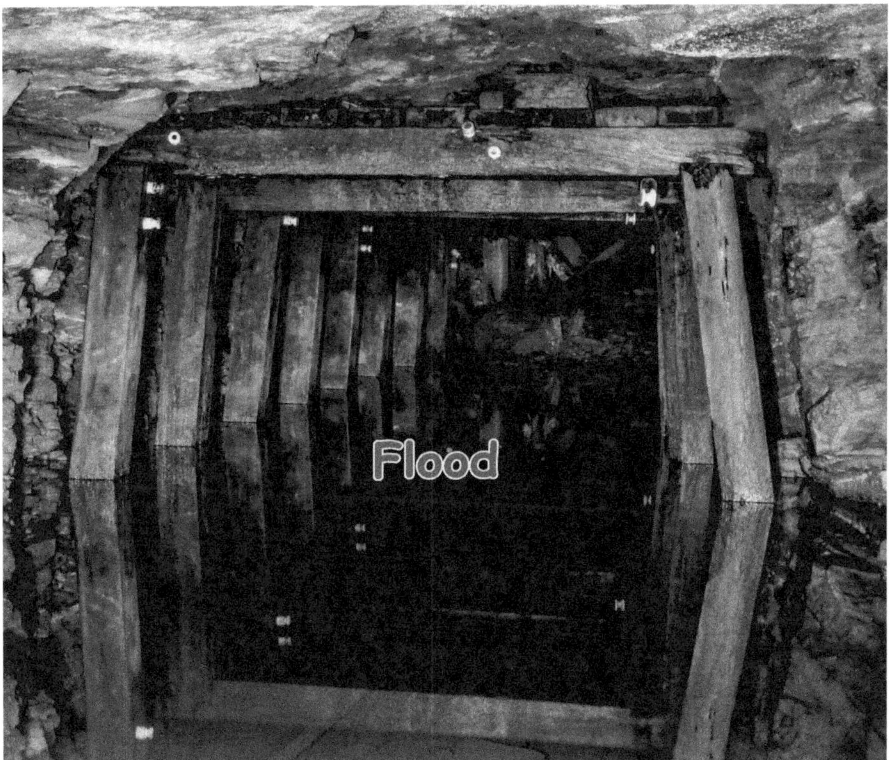

Flood

FOUR, **Steam Pumps**

Thomas Newcomen comes
up with an idea
to fix the flooded mines.

Year 1712 AD

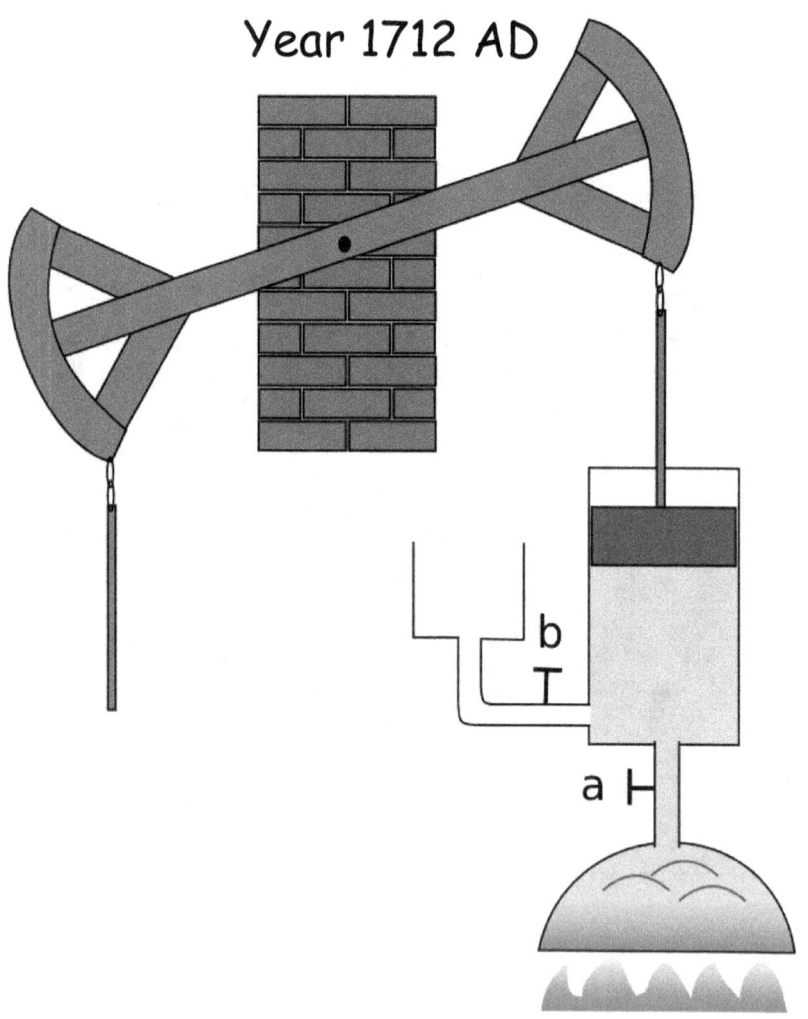

4-1

He burns coal in a beer
brewer's boiler to get steam.

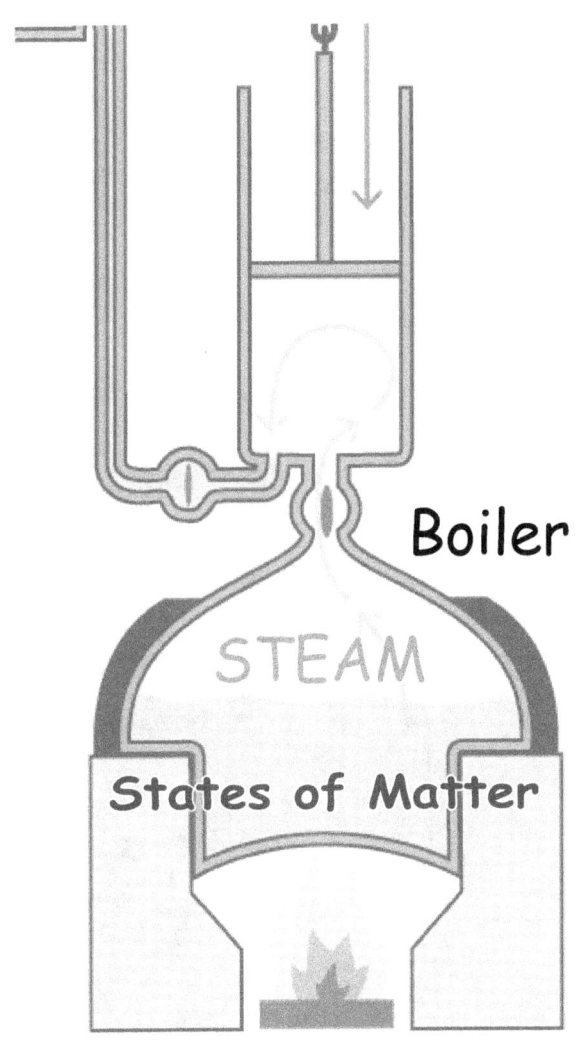

Boiler

STEAM

States of Matter

He uses the steam to push
a piston that moves up
one side of the steam pump.

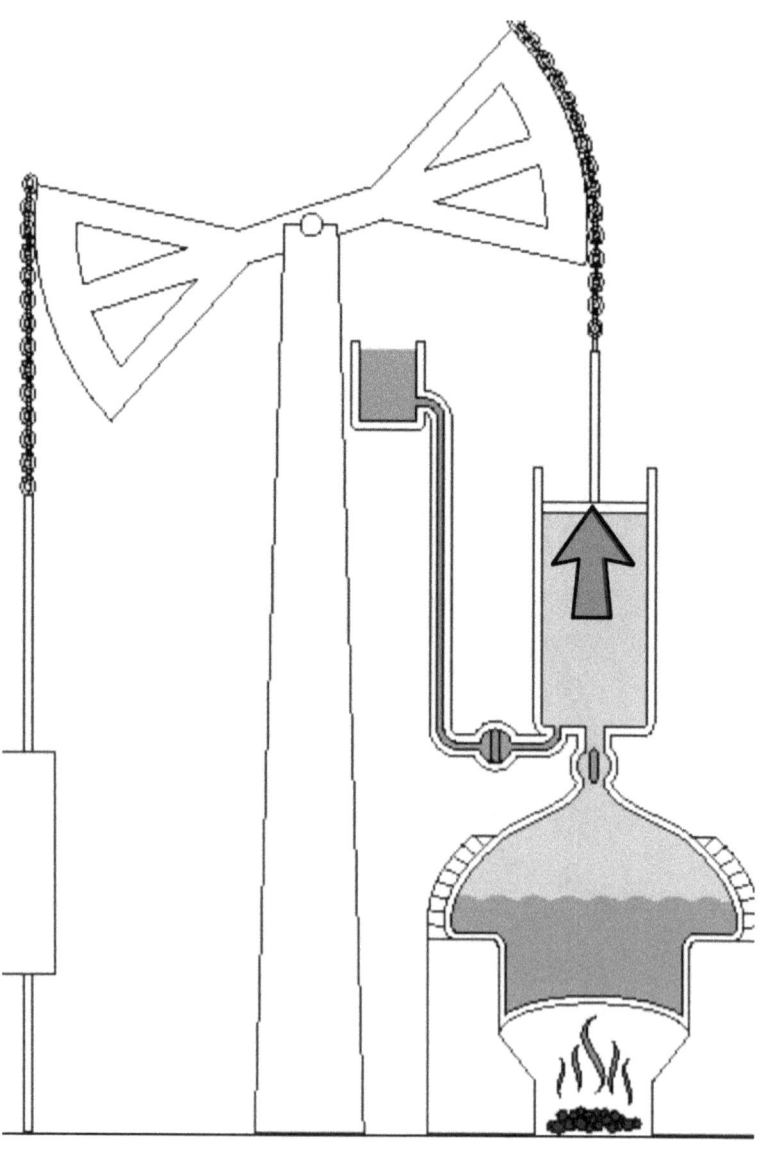

4-3

Cold water sprays into the piston and makes a vacuum. This pulls the other side up to pump water out of mines.

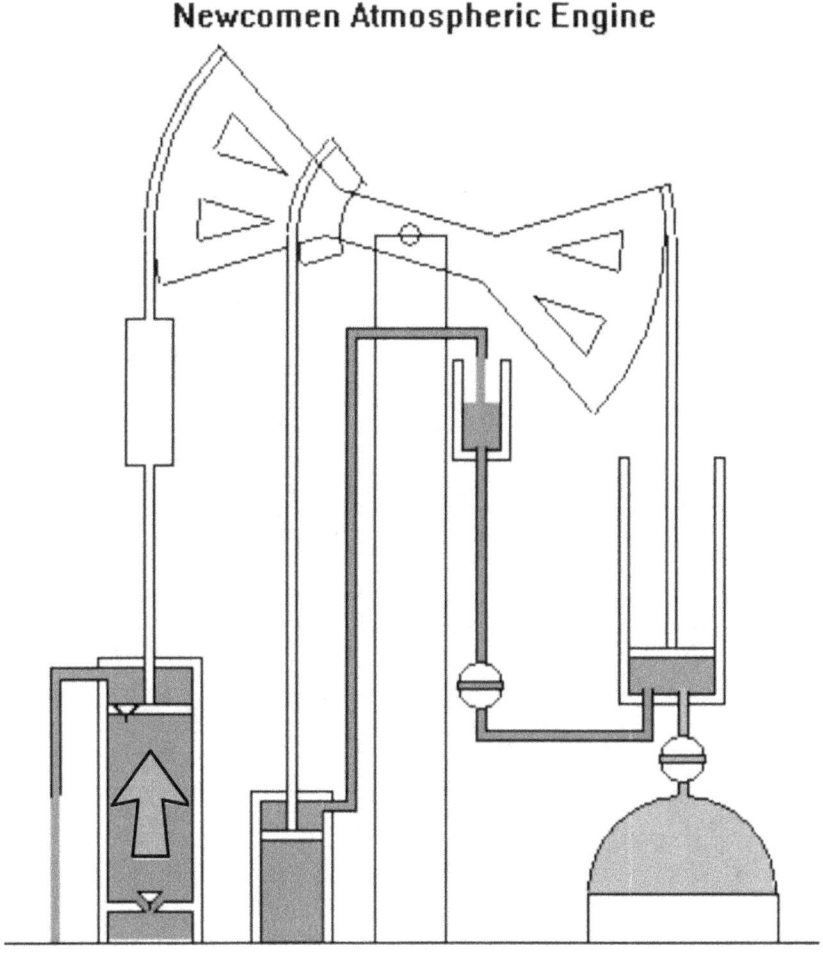

Newcomen Atmospheric Engine

The Steam Pump uses a back-and-forth see-saw motion. The pump works OK but burns lots of coal, so it is inefficient and expensive to use.

The steam pump sells pretty well. Many mine owners buy them.

But there is a problem. Pump cylinders are made out of expensive brass. A cheaper and stronger material is needed.

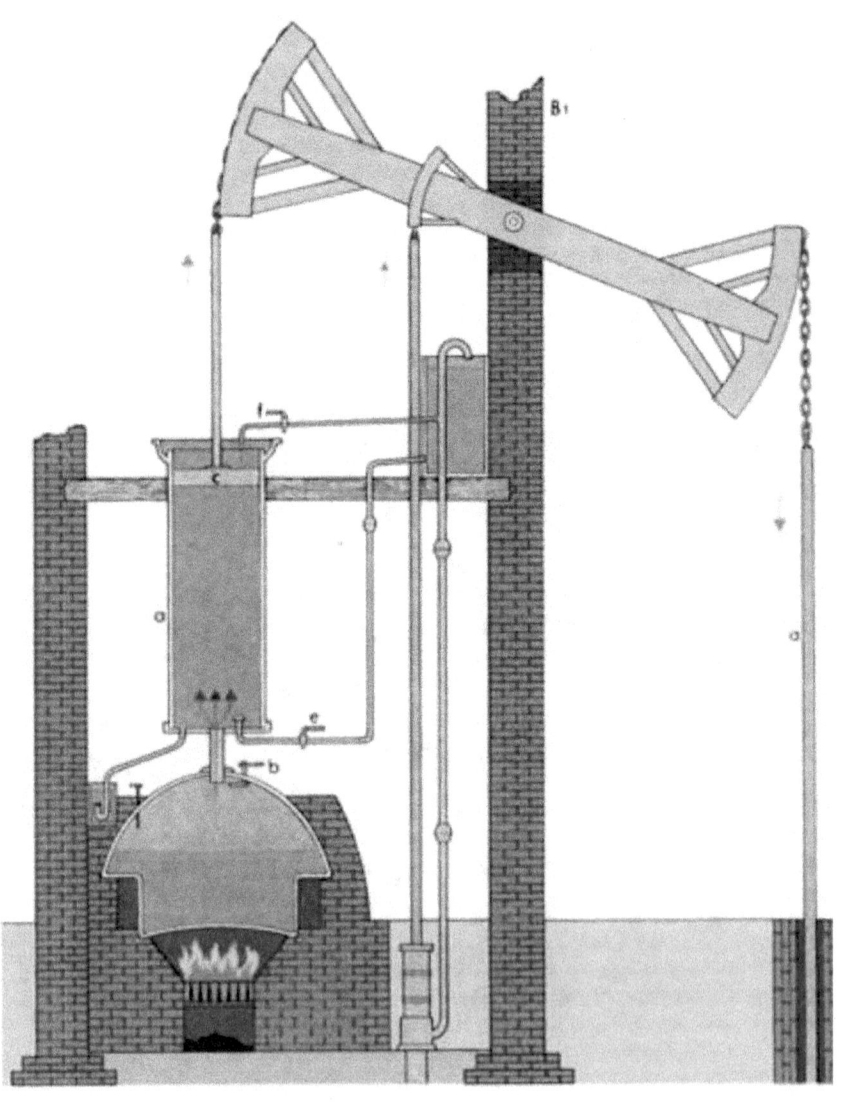

FIVE, Iron Products

At first, Abraham Darby makes brass cooking pots. Next, he discovers how to turn coal into cleaner coke. He uses coke to fuel giant blast furnaces to make iron products.

Coke

Year 1709

Smelt

He makes lots of cooking pots and other things out of iron.

As a side note, coke is made by heating coal without oxygen. It drives off impurities like tar and sulfur. It concentrates the combustible carbon. Coke burns hotter than coal.

Iron pots are made that cook blubber into whale oil used to light homes.

A whaler's trypot was used to try (boil) oil from whale blubber. The trypots stood with the flat sides adjacent in a row on the ship's deck in a space called a hurricane house. The hurricane house had a roof with chimneys above the pots and a brick 'duck pan' floor. Blubber was sliced into strips and placed in the pots where it was rendered into oil by fires underneath. The oil was ladled from one pot to the next and finally poured through feather pipes into barrels in the ship's hold. Meanwhile thick black smoke rolled all round and settled in a greasy film over ship and sailors.

These trypots were placed here on Marine Board land by the Lions Club of Hobart (Host) to mark the Hobart whaling tradition. The pots came from the seafaring Sward family of Kettering and they now belong to the City Council.

Coins collected are applied to Lions Club charities. 1981

He also makes boilers, cylinders and pistons for steam pumps.

For about 50 years, steam pumps
drive water out of deep mines.

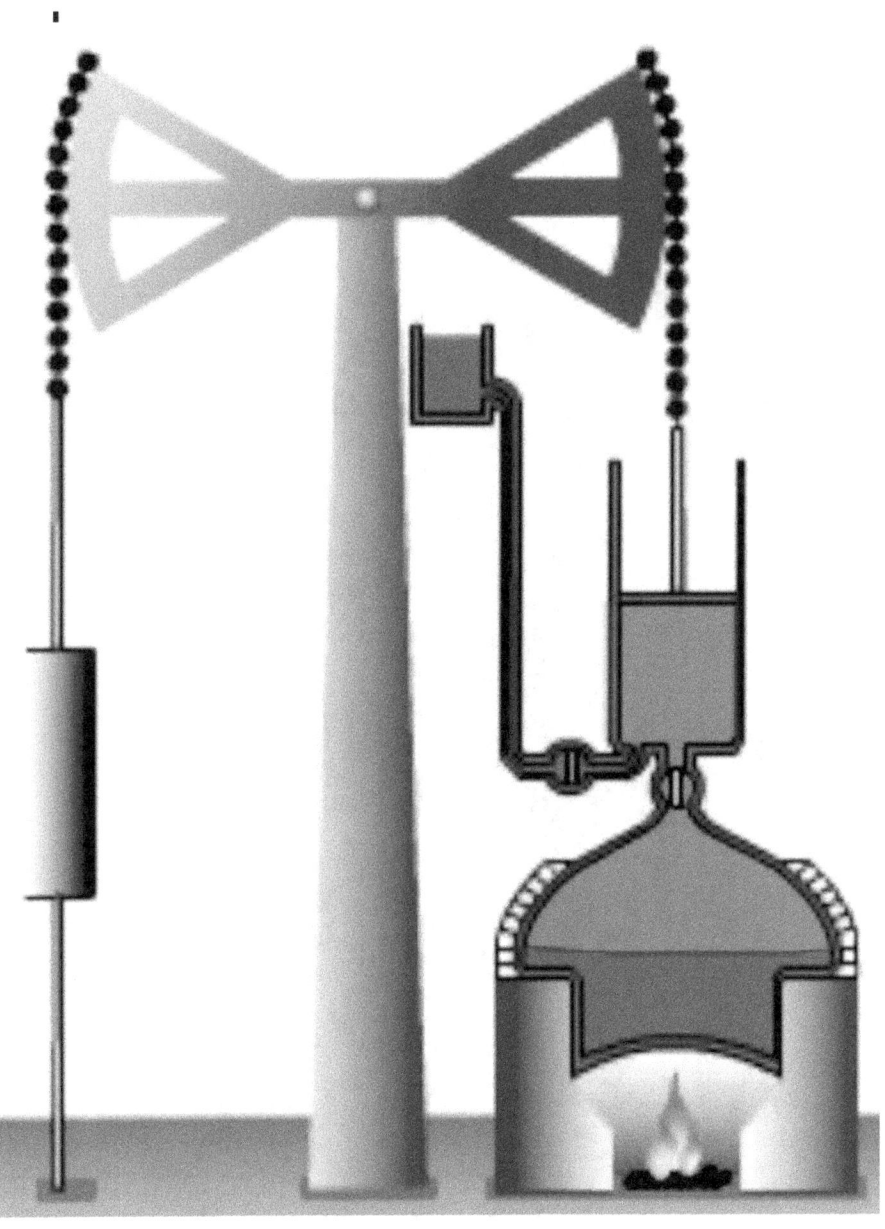

A steam pump model is used
to teach university students.
One day the model breaks.

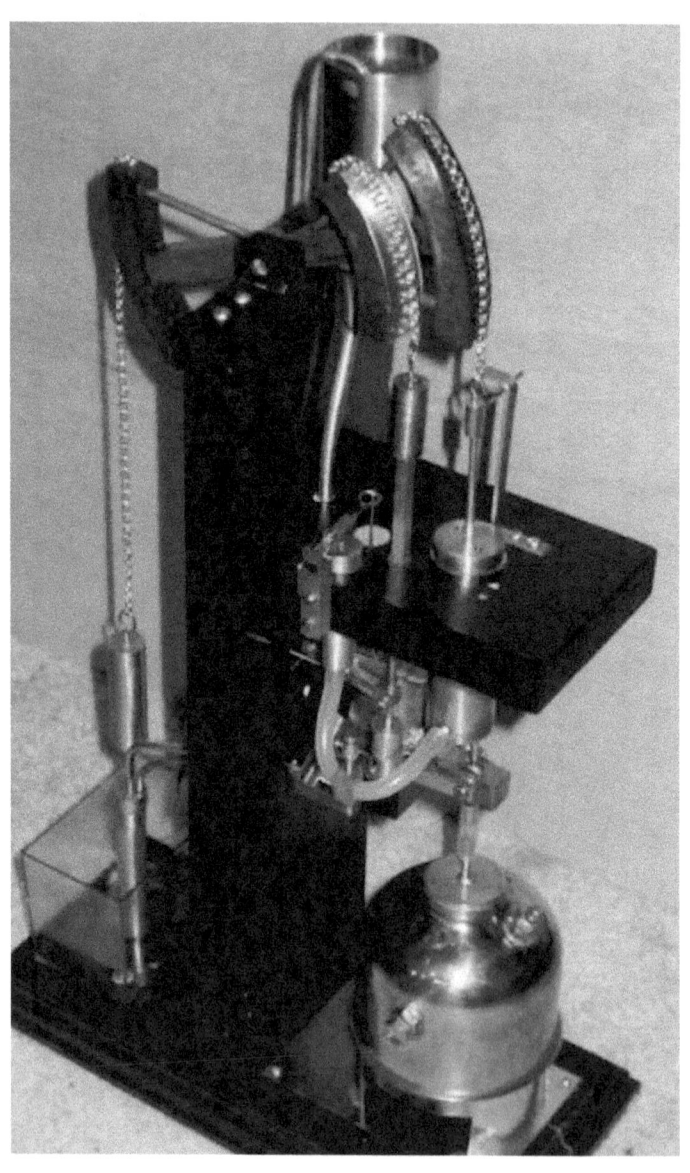

5-6

The broken model is given to the university's repairman to fix. His name is James Watt.

Steam Pump Model

University of Glasgow

SIX, Steam Power

While James Watt fixes the model, he notices something.

Steam

The steam pump uses only one cylinder for both hot and cold. This wastes a lot of energy.

Aha! Watt's brilliant idea
is to add another cylinder
to separate the hot and cold.

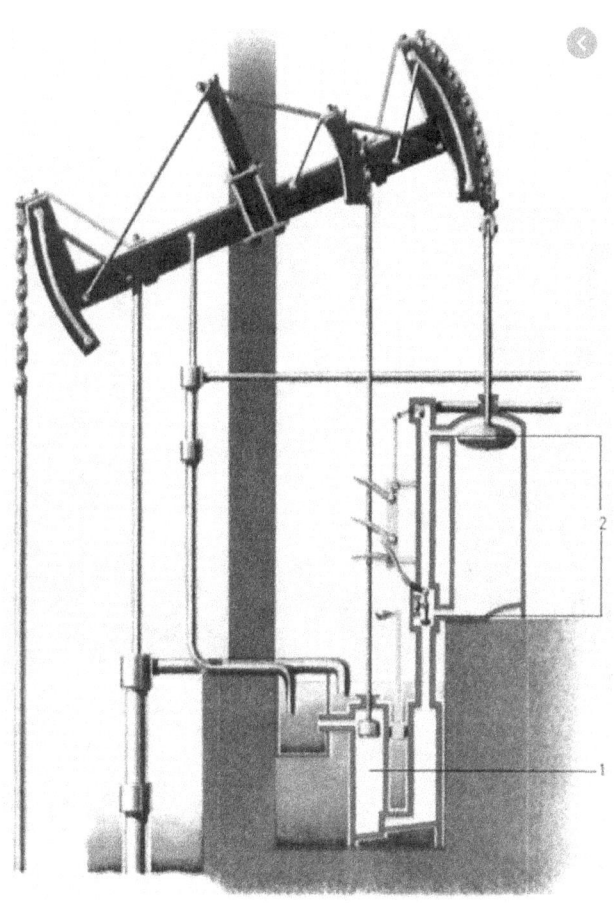

Watt's new steam engine uses 75%
less coal to push water out of mines.

His new designs cost
3/4ths less to run and are
four times more powerful
than the steam pump.

Watt works on his design
improvements for 10 years!

Watt partners with Matthew Boulton who owns a factory with engineers too.

Boulton's Factory near Birmingham, UK.

Wilkinson's Water-powered Boring Machine

A guy named "Iron Mad" Wilkinson
bores accurate, iron cylinders for steam engines.

Boulton and Watt continue to improve their **steam engines**. The biggest "wow" moment happens when the back-and-forth pistons turn round wheels.

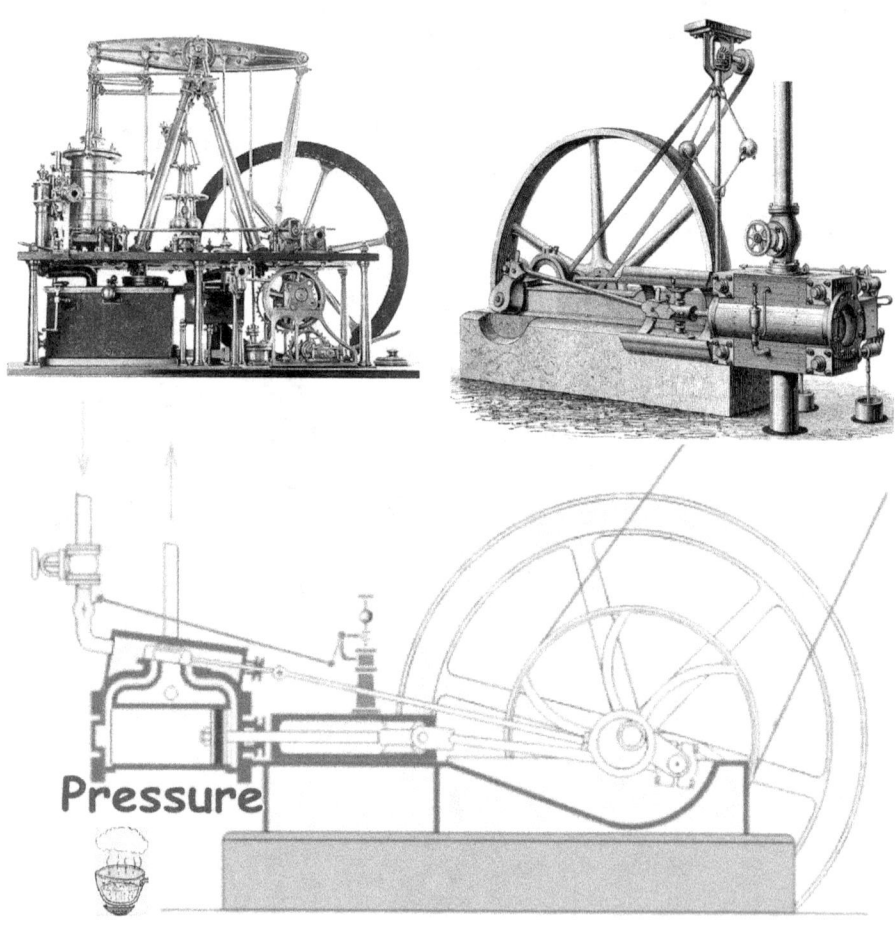

Pressure

In their first 25 years, they make 500 steam engines.

The steam engine turning motion
is a very useful power. It moves
machines for businesses like
breweries, textile makers and potteries.

Power

Steam-powered Thread Spinning Machines

At this time, steam engines still pump
water out of mines. They also
pump fresh water to people who live
in large cities as part of indoor plumbing.

Interesting side note — up to about 500 AD Romans have
indoor plumbing. After the fall of the Roman Empire, it is
over 1,000 years before Europeans have indoor plumbing again.

Steam engines power so many factories, they cause an Industrial Revolution.

Steam-powered Cloth Weaving Machines

But powerful steam engines can do more!

SEVEN, **Wheels Propel**

This guy gets the
idea to make a steam engine
to push an iron train on tracks.

Year 1804

Richard Trevithick

Paying passengers get a
short ride on a circle track
at his "Steam Circus" in London.

Year 1808

About twenty years later, there is a race to be the world's first commercial train. The place is in England on train tracks between Liverpool & Manchester.

1829 Rainhill Train Race

A train named "Rocket" wins the race.

Acceleration

George Stephenson's Train Named Rocket

Next, trains are very successful.
Soon trains and tracks spread
throughout England and Europe.

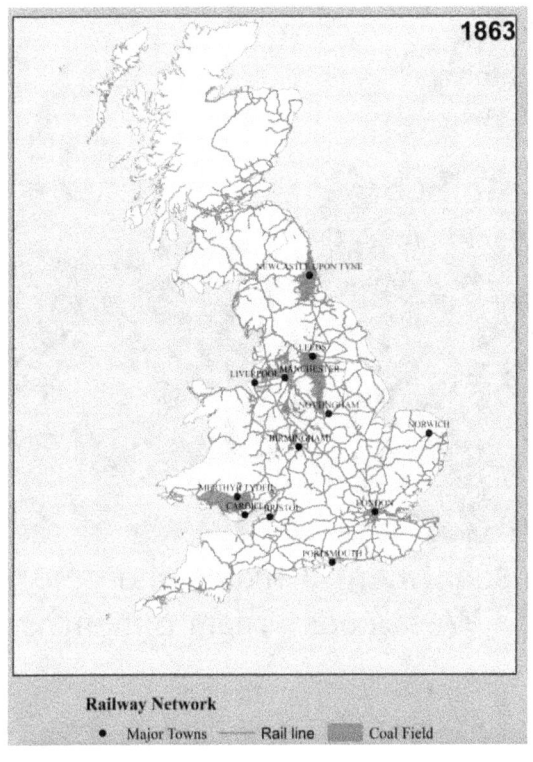

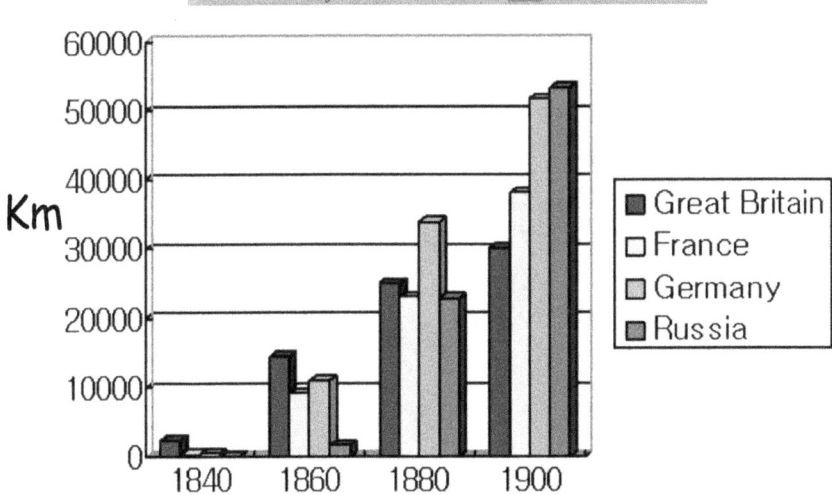

Later, steam engine trains spread across the world.

USA transcontinental trains shorten trips from the East to West coast from 6 months to 6 days.

The Trans-Siberian Railway is 10,000 kilometers long.

Steam also powers huge ships...

and some of the first cars.

Stanley Steamer

Nowadays, other engine
types power our
modern people-movers like:

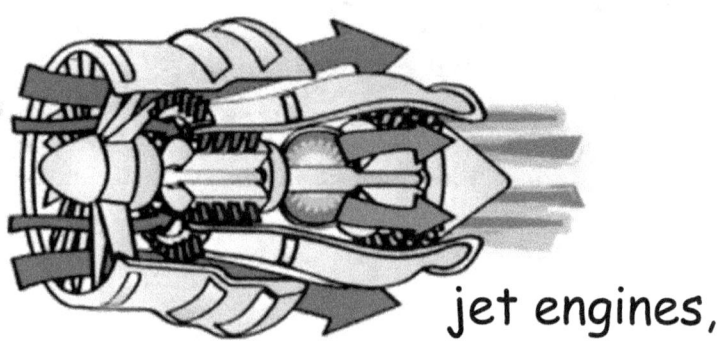

jet engines,

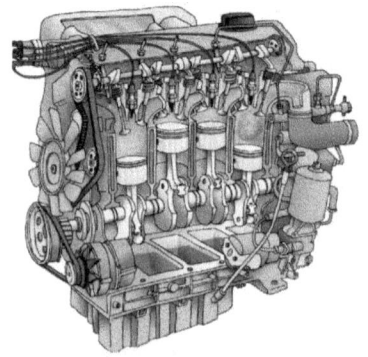

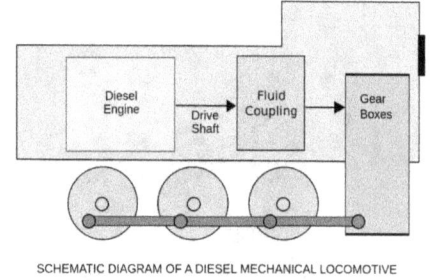

SCHEMATIC DIAGRAM OF A DIESEL MECHANICAL LOCOMOTIVE

car petrol
engines

& train diesel
engines.

But steam is still
important to modern living.

Steam still makes our coffee,
cooks much of our food
and helps iron our clothes.

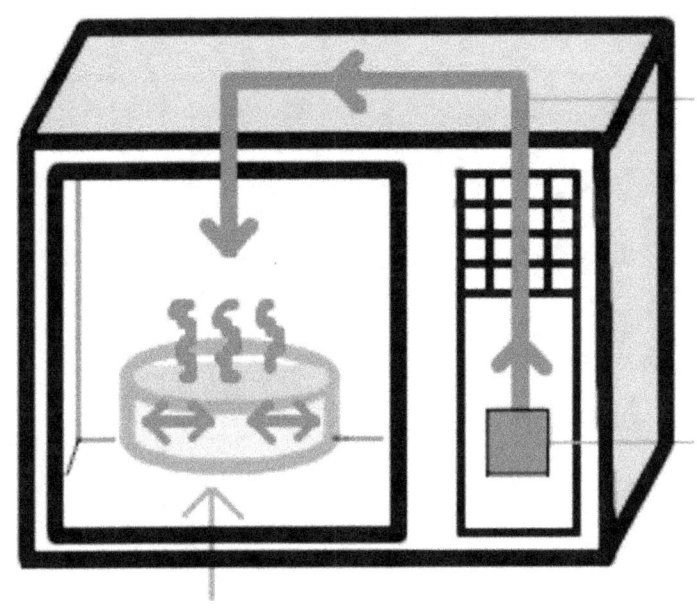

In nature, sunshine heats water
that turns into steam and evaporates.

This is part of the water cycle.

Today, steam generates much of the world's electricity too.

Steam turns turbines.

This includes coal or nuclear power that makes the steam that turns the turbines to generate our electricity.

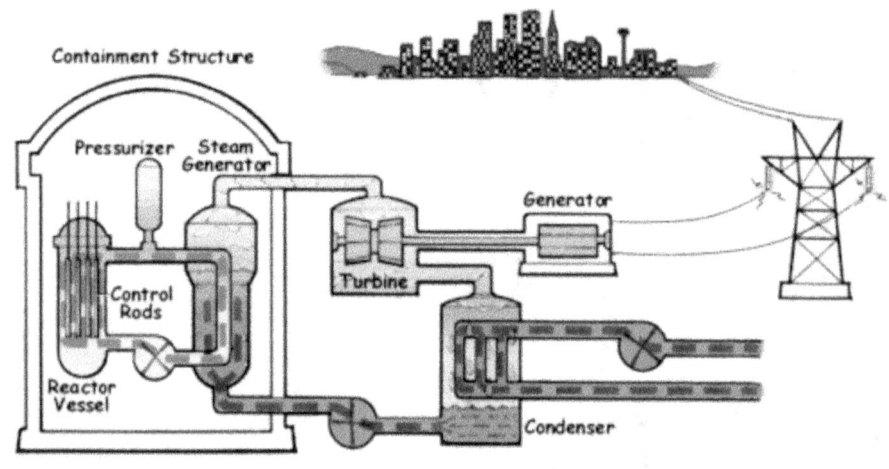

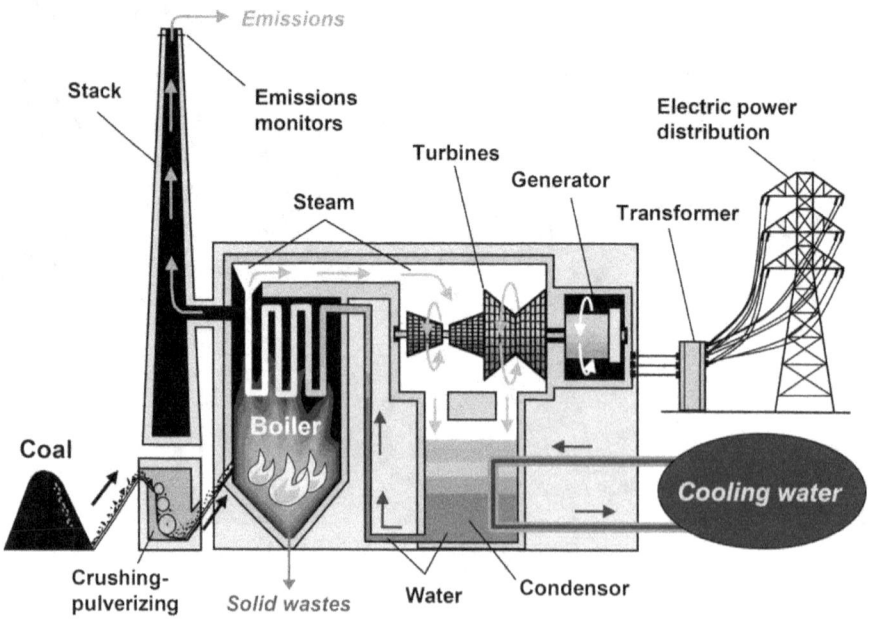

To Close

Glass windows are once made
by people with blow pipes.

This includes lots of heat to melt the glass.

Hot glass workers
drink beer to keep cool.

Power

Brewers make beer in
steaming hot copper boilers.

Vat

Steam pumps use
boilers to push water
out of flooded mines.

Next, steam engines power farms and factories.

Machines make it possible to farm with fewer workers. Many people move to cities to work in the new steam-powered factories.

In the past, steam boilers on wheels push trains around the world.

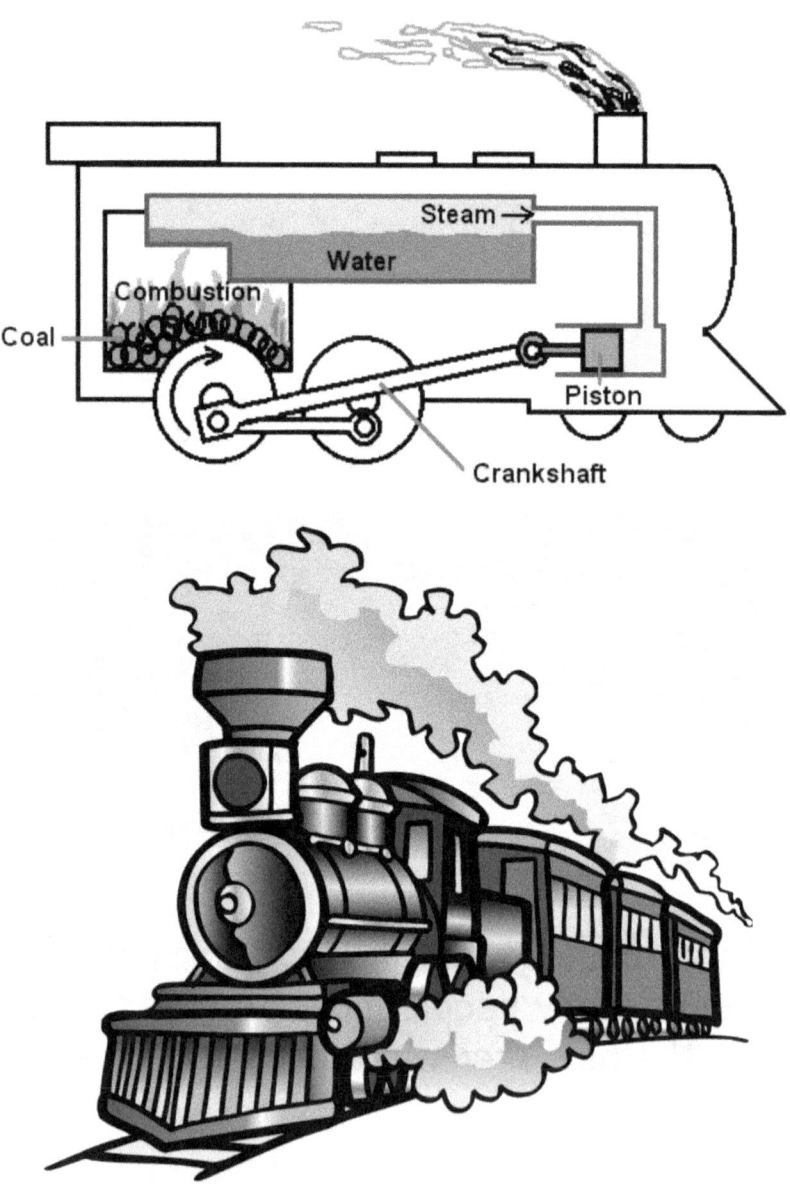

Today, steam still makes
much of our electricity.

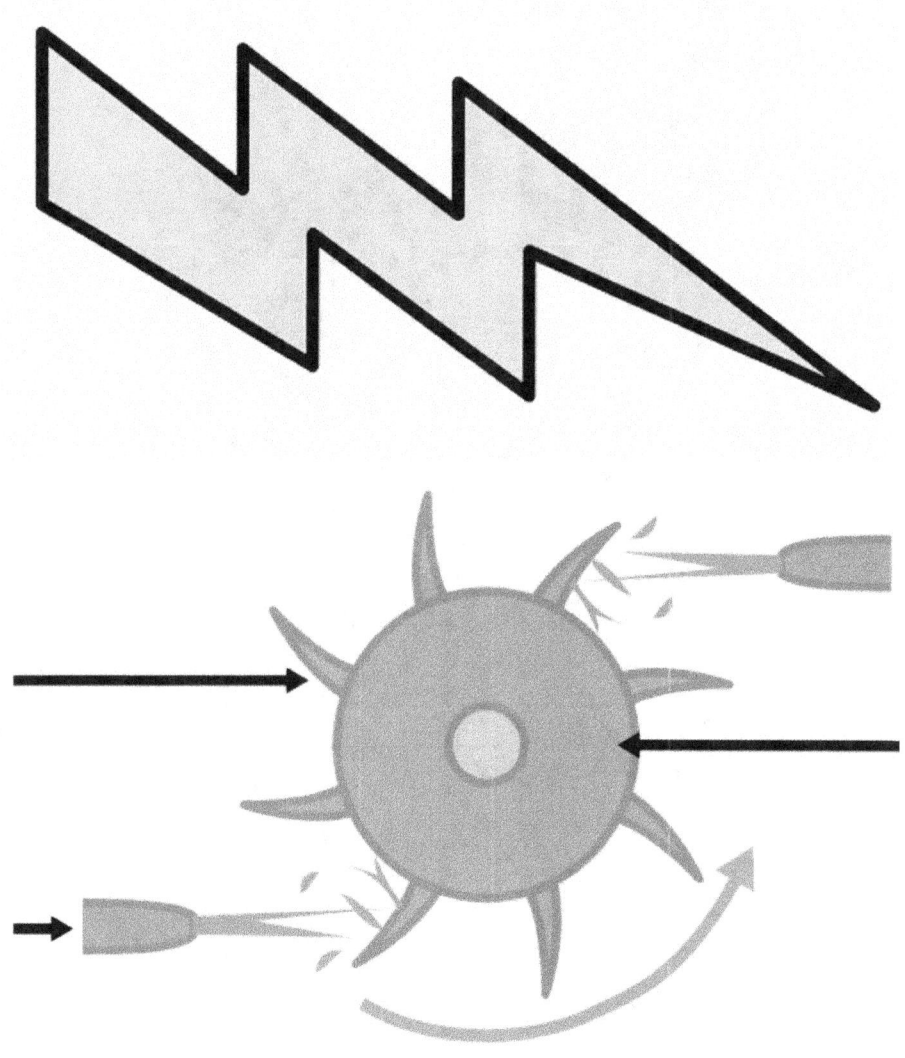

Glass, boilers and steam connect through time with power from windows to wheels.

Heat changes water into steam. It expands to power factories around the world.

C-8

POWER, One Pager

Tech connects with power from clear windows to moving wheels. In between is beer, fires and then the story gets in hot water with "steam."

For a copy of this video contact:

Welcome to "Science from Farms to Factories."In the past, most people have only just enough to eat to stay alive. Science improves our lives in 7 steps.

For a copy of this video contact:

Power
— Windows to Wheels

Long ago, workers drink beer to quench thirst at the hot, smoky glass window factory. Beer is made in giant copper vats. Someone uses the vat to make a steam pump to move water out of coal mines. Later, someone else repairs a steam pump model and gets the idea to invent the steam engine. Other people use the steam engine to power trains that have steel wheels and tracks. Science connects from windows to wheels.

What Is It?

The STEM-Zen Program

is an integrated SCIENCE Program with thousands of pages and over 50 videos. Teachers help students go from science empty to knowledge enLighted!

STEM-Zen Program Strategy
Integrated Science Curriculum

STEM-Zen
Program Guide

STEM-Zen PREP

1) Cookie Come Froms
2) Cozy Clozy
3) Good Food Goes Bad
4) No Plants No Food
5) Plants Give
6) Sand Sea
7) Senses
8) Sun Above Clouds
9) Tad's Tale
10) Too Much Tech
11) Tree Trips
12) Wing Ways

Bonus

. Cats & Dinos
. Desert Rain

. Math

— Numbers, Money, Shapes

. Home Stars
. Teams

STEM-Zen ONE

1) Seven Ideas
2) AIRPLANES
3) CARS
4) COMPUTERS
5) Smartphone-7 Waves
6) Smartphone
— Objects Before Apps
7) Electricity
8) Everyday Objects
9) Stress Less
10) VIDEO GAMES

STEM-Zen ONE
Teachers Guide

STEM-Zen TWO

1) Science Thinks
2) PLANES
— Past & Present
3) WEDGE TOOLS
—Axes to Airplanes
4) Manu-FACTORY
— Clay to Cars
5) COMPUTERS
— Then & Now
6) NETWORKS
— Wires to WiFi
7) PANDEMICS
— Causes & Cures
8) MOON RACE
— Chase to Space
9) Science of
Lucky Stars

STEM-Zen THREE

1) Air, Water & Food
2) BOTS
— Automata to AI
3) POWER
— Windows to Wheels
4) NATURE
— Where Life Lives
5) SEASONS
— Turn, Tilt & Orbit
6) Images in Action
— Why Movies Move
7) LIGHT
— Sun to Screens
8) Bright Reading
— Baas to Books

Science
by Subject

www.ingramcontent.com/pod-product-compliance
Lightning Source LLC
Chambersburg PA
CBHW072142290526
45794CB00004B/1391